SpringerBriefs in Climate Studies

SpringerBriefs in Climate Studies present concise summaries of cutting-edge research and practical applications. The series focuses on interdisciplinary aspects of Climate Science, including regional climate, climate monitoring and modeling, palaeoclimatology, as well as vulnerability, mitigation and adaptation to climate change. Featuring compact volumes of 50 to 125 pages (approx. 20,000–70,000 words), the series covers a range of content from professional to academic such as: a timely reports of state-of-the art analytical techniques, literature reviews, in-depth case studies, bridges between new research results, snapshots of hot and/or emerging topics Author Benefits: SpringerBriefs in Climate Studies allow authors to present their ideas and readers to absorb them with minimal time investment. Books in this series will be published as part of Springer's eBook collection, with millions of users worldwide. In addition, Briefs will be available for individual print and electronic purchase. SpringerBriefs books are characterized by fast, global electronic dissemination and standard publishing contracts. Books in the program will benefit from easy-to-use manuscript preparation and formatting guidelines, and expedited production schedules. Both solicited and unsolicited manuscripts are considered for publication in this series. Projects will be submitted to editorial review by editorial advisory boards and/or publishing editors. For a proposal document please contact the Publisher.

Rahat Sabyrbekov · Indra Overland ·
Roman Vakulchuk
Editors

Climate Change in Central Asia

Decarbonization, Energy Transition
and Climate Policy

Editors
Rahat Sabyrbekov
Organization for Security and Cooperation
in Europe (OSCE) Academy in Bishkek
Bishkek, Kyrgyzstan

Indra Overland
Norwegian Institute of International
Affairs (NUPI)
Oslo, Norway

Roman Vakulchuk
Norwegian Institute of International
Affairs (NUPI)
Oslo, Norway

ISSN 2213-784X   ISSN 2213-7858 (electronic)
SpringerBriefs in Climate Studies
ISBN 978-3-031-29830-1   ISBN 978-3-031-29831-8 (eBook)
https://doi.org/10.1007/978-3-031-29831-8

© The Editor(s) (if applicable) and The Author(s) 2023. This book is an open access publication.
Open Access This book is licensed under the terms of the Creative Commons Attribution 4.0 International License (http://creativecommons.org/licenses/by/4.0/), which permits use, sharing, adaptation, distribution and reproduction in any medium or format, as long as you give appropriate credit to the original author(s) and the source, provide a link to the Creative Commons license and indicate if changes were made.
The images or other third party material in this book are included in the book's Creative Commons license, unless indicated otherwise in a credit line to the material. If material is not included in the book's Creative Commons license and your intended use is not permitted by statutory regulation or exceeds the permitted use, you will need to obtain permission directly from the copyright holder.
The use of general descriptive names, registered names, trademarks, service marks, etc. in this publication does not imply, even in the absence of a specific statement, that such names are exempt from the relevant protective laws and regulations and therefore free for general use.
The publisher, the authors, and the editors are safe to assume that the advice and information in this book are believed to be true and accurate at the date of publication. Neither the publisher nor the authors or the editors give a warranty, expressed or implied, with respect to the material contained herein or for any errors or omissions that may have been made. The publisher remains neutral with regard to jurisdictional claims in published maps and institutional affiliations.

This Springer imprint is published by the registered company Springer Nature Switzerland AG
The registered company address is: Gewerbestrasse 11, 6330 Cham, Switzerland

Contents

Introduction to Climate Change in Central Asia 1
Rahat Sabyrbekov, Indra Overland, and Roman Vakulchuk

Climate Research on Central Asia: State of the Art

Climate Change: A Growing Threat for Central Asia 15
Anne Sophie Daloz

**Climate Change Science and Policy in Central Asia: Current
Situation and Future Perspectives** 23
Alisher Mirzabaev

Central Asian Decarbonisation Pathways and Carbon Pricing

**Central Asian Climate Policy Pledges Under the Paris Agreement:
Can They Be Fulfilled?** 35
Rahat Sabyrbekov, Indra Overland, and Roman Vakulchuk

**Decarbonisation Opportunities and Emerging Carbon Pricing
Instruments in Central Asia** 51
Gulim Abdi, Nurkhat Zhakiyev, and Shynar Toilybayeva

Energy Transition in Central Asia

Energy Transition in Central Asia: A Systematic Literature Review 69
Burulcha Sulaimanova, Indra Overland, Rahat Sabyrbekov,
and Roman Vakulchuk

A 'Steppe' into the Void: Central Asia in the Post-oil World 83
Morena Skalamera

**Towards a Geoeconomics of Energy Transition in Central Asia's
Hydrocarbon-Producing Countries** 95
Yana Zabanova

Local Climate Change Impacts and Adaptation in Central Asia

The Dual Relationship Between Human Mobility and Climate Change in Central Asia: Tackling the Vulnerability of Mobility Infrastructure and Transport-Related Environmental Issues 111
Suzy Blondin

A Gendered Approach to Understanding Climate Change Impacts in Rural Kyrgyzstan 123
Karina Standal, Anne Sophie Daloz, and Elena Kim

Climate Change Awareness, Norms and Stakeholders in Central Asia

The Institutionalisation of Environmentalism in Central Asia 137
Filippo Costa Buranelli

The Importance of Boosting Societal Resilience in the Fight Against Climate Change in Central Asia 149
Fabienne Bossuyt

The Culture of Recycling, Re-use and Reduction: Eco-Activism and Entrepreneurship in Central Asia 161
Aliya Tskhay

Editors and Contributors

About the Editors

Rahat Sabyrbekov is an environmental economist who specializes in decarbonization, climate change and energy transition. Rahat is a Postdoctoral Fellow at OSCE Academy and Visiting Fellow at Davis Center at Harvard University. He obtained his PhD from School of Economics and Business at Norwegian University of Life Sciences. He received his Master's degree from University of Birmingham, the United Kingdom. He teaches Economics of Sustainable Management of Mineral Resources course at the OSCE Academy. His recent publications include *Putting the Foot Down: Accelerating EV Uptake in Kyrgyzstan* (2023), *Know Your Opponent: Which Countries might Fight the European Carbon Border Adjustment Mechanism?* (2022), *Fossil Fuels in Central Asia: Trends and Energy Transition Risks* (2022).

Indra Overland is Research Professor and Head of the Research Group on Climate and Energy at the Norwegian Institute of International Affairs (NUPI). He previously headed the Russia and Eurasia Research Group at NUPI and has worked on Central Asia since 2001. He completed his Ph.D. at the University of Cambridge, followed by a three-year post-doctoral project on Central Asia and the South Caucasus. He has carried out fieldwork in all Central Asian states and has been responsible for cooperation between the OSCE Academy in Kyrgyzstan and NUPI since 2007. Every year he teaches M.A. students from all Central Asian countries on energy issues and hosts 4-5 Central Asian students in Norway. He is (co)author of *Caspian Energy Politics* (Routledge), *China's Belt and Road Initiative through the Lens of Central Asia* (Routledge), *Kazakhstan: Civil Society and Natural Resource Policy in Kazakhstan* (Palgrave) and *Renewable Energy Policies of the Central Asian Countries* (CADGAT). He is a contributing author to the IPCC's Sixth Assessment Report (Working Group II).

Roman Vakulchuk is Senior Researcher at the Norwegian Institute of International Affairs (NUPI) in Oslo and holds a Ph.D. degree in Economics from Jacobs University

Bremen in Germany. He specializes in Central Asia and Southeast Asia and his main research interests are economic transition, trade, energy, climate change and investment policy. Vakulchuk has served as project leader in research projects organized by the Asian Development Bank (ADB), the World Bank, the Global Development Network (GDN), the Natural Resource Governance Institute (NRGI) and others. In 2018, he worked as governance expert for OECD's mission in Kazakhstan and advised the government on privatization reform. In 2013, Vakulchuk was awarded the Gabriel Al-Salem International Award for Excellence in Consulting. Recent publications include *Seizing the Momentum. EU Green Energy Diplomacy towards Kazakhstan* (2021), *Discovering Opportunities in the Pandemic? Four Economic Response Scenarios for Central Asia* (2020) and *Renewable Energy and Geopolitics: A Review* (2020).

Contributors

Gulim Abdi Astana IT University, Astana, Kazakhstan

Suzy Blondin University of Neuchâtel, Neuchâtel, Switzerland

Fabienne Bossuyt Department of Political Science, Ghent University, Ghent, Belgium

Filippo Costa Buranelli University of St Andrews, St Andrews, Scotland, UK

Anne Sophie Daloz CICERO, Oslo, Norway

Elena Kim American University of Central Asia, Bishkek, Kyrgyzstan

Alisher Mirzabaev Institute of Food and Resource Economics (ILR), University of Bonn, Bonn, Germany

Indra Overland Norwegian Institute of International Affairs (NUPI), Oslo, Norway

Rahat Sabyrbekov Organization for Security and Cooperation in Europe (OSCE) Academy in Bishkek, Bishkek, Kyrgyzstan

Morena Skalamera Leiden University, Leiden, The Netherlands; Belfer Center, Harvard Kennedy School, Cambridge, MA, USA

Karina Standal CICERO, Oslo, Norway

Burulcha Sulaimanova Organization for Security and Cooperation in Europe (OSCE) Academy, Bishkek, Kyrgyzstan

Shynar Toilybayeva CAREC Country Office, Astana, Kazakhstan

Aliya Tskhay University of St Andrews, St Andrews, Scotland, UK

Roman Vakulchuk Norwegian Institute of International Affairs (NUPI), Oslo, Norway

Yana Zabanova Research Institute for Sustainability (RIFS)—Helmholtz Centre, Potsdam, Germany

Nurkhat Zhakiyev Astana IT University, Astana, Kazakhstan

Introduction to Climate Change in Central Asia

Rahat Sabyrbekov, Indra Overland, and Roman Vakulchuk

Abstract This chapter provides a broad introduction to the impact of climate change in Central Asia, a region that has been experiencing a greater rise in temperatures than other parts of the world. The chapter shows how climate change represents a significant threat to Central Asia, exacerbating existing economic and environmental challenges and fueling regional tensions over resource management. Inefficient water resource management at the national level and limited regional collaboration on the management of water resources, coupled with state capacities that remain insufficient to tackle climate change impacts, compound water-related tensions between the countries in the region. The chapter also shows how decarbonisation efforts in Central Asia are still in their early stages, with coal remaining a primary source of energy. Although the Central Asian countries have announced decarbonisation targets and adopted green economy strategies and programmes to reduce greenhouse gas emissions, a large-scale clean energy transition remains unlikely in the short term. The chapter concludes by identifying a lack of scholarship on climate change in Central Asia, which limits the development of a coherent approach to climate change mitigation and adaptation and evidence-based decision-making in the region. The chapter argues that a more coordinated approach to tackling climate change across the region is needed, requiring closer collaboration and more effective joint management of natural resources by the five Central Asian states. Finally, the chapter presents the chapters in the rest of the book.

Keywords Climate change · Central Asia · Water resources · Decarbonisation · Knowledge gap

---

R. Sabyrbekov (✉)
Organization for Security and Cooperation in Europe (OSCE) Academy, Bishkek, Kyrgyzstan
e-mail: r.sabyrbekov@osce-academy.net

I. Overland · R. Vakulchuk
Norwegian Institute of International Affairs (NUPI), Oslo, Norway
e-mail: ino@nupi.no

R. Vakulchuk
e-mail: rva@nupi.no

1 The Impact of Climate Change on Central Asia: What Is at Stake?

Central Asia will be heavily impacted by climate change. In its Sixth Assessment Report, the Intergovernmental Panel on Climate Change (IPCC) concluded that climate change presents a major threat in Central Asia, where temperatures have been rising more than the global average (IPCC 2022). Climate change has the potential to exacerbate existing economic and environmental challenges and fuel regional tensions over common resource management, particularly as Central Asia remains a politically and economically unintegrated region (Vakulchuk and Overland 2019).

Climate change will compound existing water-related tensions between Kyrgyzstan, Tajikistan and Uzbekistan. Central Asia has long suffered from inefficient water resource management at the national level and sluggish regional collaboration in the management of the limited water resources. Individual state capacities are also limited and remain insufficient to tackle climate change impacts. Furthermore, Tajikistan and Kyrgyzstan are highly dependent on agriculture, the sector which is most vulnerable to climate change, while oil and gas exporters Kazakhstan and Turkmenistan may be impacted by looming carbon taxes in export markets and the risk that their fossil fuel could become stranded assets. The countries' energy mix and contribution to the global GHG emissions (see Fig. 1). The Central Asian countries have declared ambitious climate goals but continue to rely heavily on fossil fuels and outdated, crumbling energy infrastructure.

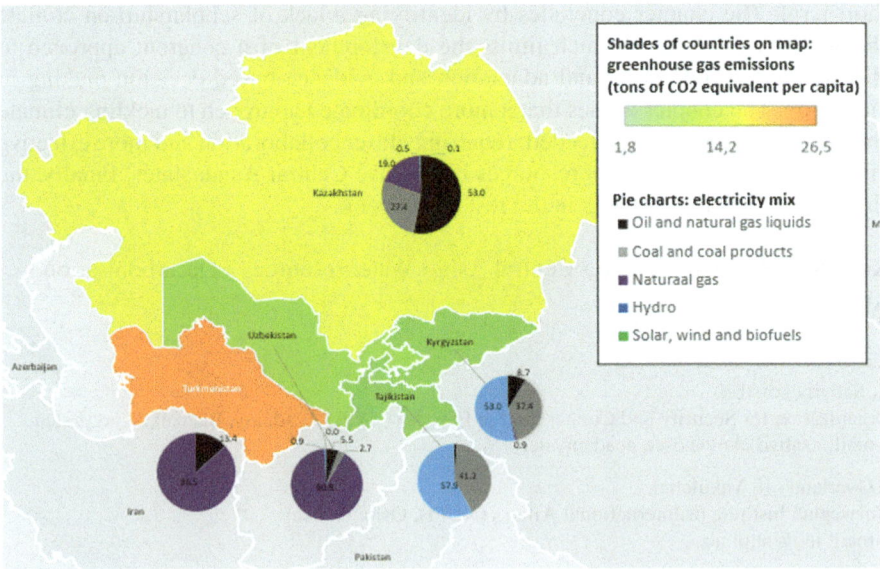

Fig. 1 CO_2 equivalent emissions per capita and energy mix in Central Asia

Effective climate policy will require closer collaboration and more effective joint management of natural resources by the five Central Asian states. The current lack of scholarship on climate change in Central Asia limits efforts to fully understand the issues within each country and across the region, as well as the development of a coherent approach to climate change mitigation and adaptation. This volume represents an attempt to close some of the gaps in the scholarship, stimulate cross-country comparison, and identify both country-specific and shared challenges. We hope this will be a useful first step towards developing a more coordinated approach to tackling climate change across the region.

2 Decarbonisation in Central Asia

For several decades, Central Asia has been a source of fossil fuels for the world economy, mainly in the form of oil and gas (Palazuelos and Fernández 2012; Vakulchuk 2016). This has made the region highly dependent on revenues from hydrocarbon exports (Vakulchuk and Overland 2021). Oil and gas revenues accounted for 35% of GDP and 75% of exports from Kazakhstan in 2020 (IMF 2021). Natural gas exports to China alone accounted for 25% of Turkmenistan's GDP in 2020 (CIA 2021), while natural gas production made up 15% of Uzbekistan's GDP in 2018. Consumption of fossil fuels by the region itself has also been growing since 2010 (Vakulchuk et al. 2022b). Coal is one of the main energy sources in Central Asia, and its consumption is on the rise in Kyrgyzstan and Tajikistan. Coal heating was one of the main reasons why the Kyrgyz capital, Bishkek, was the world's most polluted city in 2021 (World Air Quality 2021; Sabyrbekov and Overland 2020). Air pollution at this level is a severe health hazard.

All Central Asian countries are signatories to the Paris Agreement and have announced decarbonisation targets. Since 2018, the Central Asian states have also been promoting renewable energy. Most of the region's countries have adopted at least some type of green economy strategy or programme to increase resource use efficiency and reduce greenhouse gas (GHG) emissions. The results of these programmes are mixed. The share of renewables in power generation in Kazakhstan was only 2.5% in 2020, while that of coal was 70% (IEA 2021), and the share of solar and wind power in the energy mix remains negligible in all countries. A serious clean energy transition remains unlikely in the short term (Sabyrbekov and Ukueva 2019; Vakulchuk et al. 2022b).

3 Climate Change in Central Asia: Status of Knowledge

Since the 2010s, the IPCC has expressed concern over the fact that Central Asia is the least studied region with respect to climate change despite the region's vulnerability to its impacts (IPCC 2014). Out of 54 thematic areas critical to the impacts of climate change, 51 remain underexplored in Central Asia, with severe knowledge gaps or no data available at all in these areas.

In response to the IPCC's observation, Vakulchuk et al. (2022a) conducted a systematic review of the scholarship on climate change in Central Asia in the natural and social sciences over the past thirty years to ascertain the extent to which the academic community has engaged with this increasingly urgent multi-layered issue. The authors found that, out of 1,305 conference panels organised by international Central Asia studies associations between 1991 and 2020, none focused on climate change. Furthermore, out of the 10,249 individual conference presentations, only two (0.02%) were on a topic related to climate change.

There is a particular lack of scholarship on climate change in Central Asia in the social sciences. According to Vakulchuk et al. (2022a), there were half as many publications on climate-related topics in Central Asia in the social sciences as in the natural sciences between 1991 in 2022, and only a handful of researchers have a track record of publishing in both the natural and social sciences (e.g., Kerimray et al. 2015, 2018). Therefore, in terms of research specialisation, funding and institutional infrastructure, the social sciences are lagging behind the natural sciences. Moreover, the existing literature has neglected the political, economic, social, health, security and geopolitical domains. For example, to our knowledge, only two articles have been published on climate change and health in Central Asia (Janes 2010; Bhuiyan and Khan 2011), and only one on climate change and gender (Kim and Standal 2019). No publications cover the issues of climate justice, transportation or climate diplomacy in Central Asia (Vakulchuk et al. 2022a). The recent dates on most of the published works do, however, indicate that social science scholars may finally be starting to take an interest in this field of research.

4 Our Mission

To address the urgent need for scholarship on climate change in Central Asia, and in particular the need for research by social scientists, this book includes contributions from a range of contributors relating to climate change impacts, adaptation and mitigation in the region. Throughout the book, we try to capture the broader societal implications of climate-related issues. The book includes 12 chapters in addition to this introductory chapter. Each chapter makes an important contribution to social science scholarship on climate change and decarbonisation in Central Asia and covers a topic that has received little or no attention in the literature to date. Many of the chapters cover the entirety of the Central Asian region, while some focus on individual countries.

This book is important in three ways. First, taking into consideration the significant gaps in social science scholarship on climate change identified by Vakulchuk et al. (2022a), we aim to provide the first systematic contribution towards filling some of the urgent gaps in the knowledge and data. This can accelerate knowledge-building on climate change in Central Asia. Second, by inviting scholars in the field of Central

Asian studies to explore the nexus between climate change and other societal topics in Central Asia, we aim to raise awareness within the Central Asia studies community about the urgent need to integrate climate-related analysis into social science research. This, in turn, can facilitate more evidence-based decision-making, policies and projects in the areas of climate change adaptation and mitigation in Central Asia on the part of donors and the authorities in the Central Asian states. Third, through the production of new knowledge, we also seek to assist local communities in taking well-informed actions to adapt to climate change impacts.

5 Overview of Chapters

5.1 Part I: Climate Research on Central Asia: State of the Art

In her chapter 'Climate Change: A Growing Threat for Central Asia', Anne Sophie Daloz takes stock of the major physical impacts of climate change in Central Asia based on the most recent literature, including the Sixth Assessment Report of the IPCC published in 2022. She identifies climate change-related risks and sectoral vulnerabilities across the region, and her analysis serves as a starting point for the other chapters of the book. Daloz explains how climate impact-drivers are expected to change over time, thus amplifying the vulnerability of the agriculture, health, energy, and transportation sectors. She also argues that, without adequate adaptation measures, Central Asia could face severe water scarcity, which will have detrimental knock-on effects on energy and food security in the region.

In the second chapter in this part of the book, 'Climate Change Research and Policy in Central Asia: Current Situation and Future Perspectives', Alisher Mirzabaev conducts a bibliometric analysis to explore the state of scientific research related to climate change adaptation and mitigation. He assesses key trends and discusses investment priorities for research. Similarly to Vakulchuk et al. (2022a), Mirzabaev argues that, although the scientific literature on climate change in Central Asia has started growing, it remains too small and underdeveloped to grasp the complexity of climate impacts. He calls for open access to data, more investment in climate change research, the development of local climate modelling capacities, and support for regional knowledge and scientific exchange on climate change topics.

5.2 Part II: Central Asian Decarbonisation Pathways and Carbon Pricing

In the chapter 'Central Asian Climate Policy Pledges Under Paris Agreement: Can They Be Fulfilled?' we explore the region's preparedness to meet its climate policy goals under the Paris Agreement. We compare the countries' Nationally Determined

Contributions (NDCs) with their national strategic programmes and development trajectories. Our findings indicate that the Central Asian countries vary in their ability to meet their declared climate goals, and that successful implementation will require structural changes in energy systems, substantial investment in infrastructure and, most importantly, the alignment of development plans with climate goals.

In 'Decarbonisation Opportunities and Emerging Carbon Pricing Instruments in Central Asia', Gulim Abdi, Nurkhat Zhakiyev and Shynar Toilybayev discuss the region's heavy dependence on fossil fuels and its vulnerability to climate change. They identify and discuss existing and potential carbon pricing instruments in Central Asia, including emissions trading measures. The authors demonstrate that different carbon pricing mechanisms and decarbonisation strategies are being considered across the entire region. They conclude by proposing mid- and long-term activities that could strengthen regional cooperation on decarbonisation.

5.3 Part III: Energy Transition in Central Asia

In their chapter 'Energy Transition in Central Asia: A Systematic Literature Review', Burulcha Sulaimanova, Indra Overland, Rahat Sabyrbekov and Roman Vakulchuk provide a comprehensive bibliometric overview of the profile and trajectory of the research on energy in Central Asia. The authors find that research shifted from fossil fuel towards clean energy topics between 1991 and 2022. However, despite recent growth in this area, research on energy transition and the importance of renewables in Central Asia is still sparse. The chapter reveals that US and European researchers initially published most of the energy research on the region but, starting from 2016, were overtaken by scholars from China, Japan, Kazakhstan and Russia.

Morena Skalamera in her chapter 'A "Steppe" into the Void: Central Asia in the Post-oil World' delves into the nexus between the effects of energy transition, regime stability in Central Asia's fossil-fuel-dominated economies and international stability. She raises concerns about the unpreparedness of Central Asian petrostates to shift to clean energy and decarbonise their economies, as well as the risks this poses internally and to external partners.

In the chapter 'Towards a Geoeconomics of Energy Transition in Central Asia's Hydrocarbon-Producing Countries', Yana Zabanova analyses the implications of energy transition for Kazakhstan, Uzbekistan and Turkmenistan. She finds that Kazakhstan remains the region's renewable energy frontrunner, that late starter Uzbekistan has attracted major industry players, and that Turkmenistan has yet to take its first steps. Overall, the chapter finds that the three countries do not plan a fossil-fuel phaseout. Rather, they seek to add renewables to the energy mix to bolster energy security and help decarbonise their economies.

5.4 Part IV: Local Climate Change Impacts and Adaptation in Central Asia

In 'The Dual Relationship Between Human Mobility and Climate Change in Central Asia: Tackling the Vulnerability of the Mobility Infrastructure and Transport-Related Environmental Issues', Suzy Blondin addresses the impact of transportation on the region's climate through the concept of 'climate mobilities'. Old vehicles and a lack of public transport contribute to severe air pollution in Central Asian cities. Drawing on Blondin's fieldwork in Tajikistan, the chapter highlights the adverse impacts of mobility disruptions for the populations of Central Asia. After discussing the complex relationship between mobilities and climate change, Blondin addresses the interconnection between climate justice and mobility justice and concludes with policy recommendations to promote sustainable mobilities in the region.

In the chapter 'A Gendered Approach to Understanding Climate Change Impacts in Rural Kyrgyzstan', Karina Standal, Anne Sophie Daloz and Elena Kim address the lack of research on the intersectional dynamics of climate impacts and gender. The authors empirically examine the effects of climate change in rural Kyrgyzstan and chart the experiences and realities of local women, applying a contextual vulnerability lens. Standal et al. combine analysis of existing research on the physical impacts of climate change with ethnographic accounts of local people's farming and energy-use practices. This approach allows them to better understand the interlinkages between the material, social and cultural realities of local communities. The authors conclude that social differentiation both enables and limits the varied capabilities of people to cope with climate change impacts and engage in adaptation practices.

5.5 Part V: Climate Change Awareness, Norms and Stakeholders in Central Asia

In the chapter 'The Institutionalisation of Environmentalism in Central Asia', Filippo Costa Buranelli explores how the notion of environmentalism has been internalised and institutionalised in Central Asia. The chapter is a case study of the Central Asian republics taking part in the 26th Conference of the Parties to the United Framework Convention on Climate Change (COP 26) as a single entity advocating for a shared approach to climate change. Buranelli considers whether, how, and to what extent the Central Asian governments have institutionalised environmentalism in the context of climate change, focusing on state-based and people-based discourses, initiatives and campaigns at the global, regional and local levels. He concludes that the institutionalisation of environmentalism is strongest at the regional level and weakest domestically, with cooperation at the regional level developing but still dependent on international organisations and donors.

Fabienne Bossuyt's chapter, 'The Importance of Boosting Societal Resilience in the Fight Against Climate Change in Central Asia', underlines the urgent need for societies in Central Asia to learn how to adapt to the effects of climate change. Bossuyt argues that resilience strategies involving the sharing of responsibilities among individuals and communities will increase the ability of these countries to withstand the impacts of climate change. She observes that Central Asian societies have a strong tradition of home-grown solidarity movements and locally embedded self-reliance practices. The Central Asian governments and international donors could therefore help boost societal resilience to climate change in Central Asia by supporting the ability of local social actors to self-organise and utilise local strengths and knowledge of available resources and infrastructure.

In her chapter, 'The Culture of Recycling, Re-use and Reduction: Eco-Activism and Entrepreneurship in Central Asia', Aliya Tskhay focuses on the role of the private sector in addressing climate challenges by changing people's behaviour and attitudes towards the environment. She explores the benefits and challenges of private eco-activism and entrepreneurship at the micro level. Tskhay looks at private companies in Almaty that have launched initiatives to encourage the reduction of consumption and associated retail waste, along with social awareness programmes. Her chapter demonstrates how entrepreneurs and activists are filling the gap in sustainability promotion, while also revealing more profound issues with commitments to carbon reduction and other environmental projects, as well as barriers to societal readiness to embrace changes in entrenched behaviours.

6 Concluding Remarks

Clearly, climate change will have negative consequences throughout the Central Asian region, particularly given the region's exposure and vulnerability to climate change. These impacts will be felt at different levels. At the household level, the most vulnerable groups, such as women and the poor, are most exposed to climate-induced risks. At the macroeconomic level, agriculture, infrastructure and the energy supply will be heavily disrupted and will require large investments. At the regional level, some have argued that the ongoing global transition to clean energy may destabilise the political regimes of the Central Asian petrostates. Perhaps seeing this potential, the petrostates were the first among the region's countries to embrace renewable energy projects. Despite this, no phaseout of fossil fuels has thus far been planned.

The Central Asian countries vary in terms of their adaptation and mitigation plans as laid out in their climate pledges under the Paris Agreement, but what they have pledged is generally not ambitious. Decarbonisation initiatives remain in the early stages and are not aligned with the national development strategies of the countries. Furthermore, all of the Central Asian countries have made their most ambitious climate pledges conditional upon substantial financial and technical support from donors.

Despite the common threats and shared resources, the countries have been slow to take a regional approach to climate change adaptation and mitigation. Moreover, heedless of the looming threat of climate change, research has been sluggish, with those publications that are produced dominated by scholars from China, Kazakhstan and Russia. Further and broader research in the area is much needed to facilitate informed policy decisions by governments and local communities. The region's success in meeting present and future climate challenges is dependent on greater effort in terms of regional cooperation, integration of climate policy into sector programmes and the active involvement of local stakeholders at all levels.

References

Barandun M et al (2018) Multi-decadal mass balance series of three Kyrgyz glaciers inferred from modelling constrained with repeated snow line observations. Cryosphere 12(6):1899–1919

Barandun M et al (2020) The state and future of the cryosphere in Central Asia. Water Secur 11:100072

Barandun M et al (2021) Hot spots of glacier mass balance variability in Central Asia. Geophys Res Lett 48:e2020GL092084

Bhuiyan SH, Khan HT (2011) Climate change and its impacts on older adults' health in Kazakhstan. NISPAcee J Public Adm Policy 4(1):97–119

Bolch T (2007) Climate change and glacier retreat in Northern Tien Shan (Kazakhstan/Kyrgyzstan) using remote sensing data. Glob Planet Chang 56(1–2):1–12

Bolch T (2017) Asian glaciers are a reliable water source. Nature 545(7653):161–162

CIA (2021) Explore all countries—Turkmenistan. Central Asia. The World Factbook.

Hagg W et al (2006) Modelling of hydrological response to climate in glacierized Central Asian catchments. J Hydrol 332(1–2):40–53

Hagg W et al (2018) Future climate change and its impact on runoff generation from the debris-covered Inylchek glaciers, Central Tian Shan, Kyrgyzstan. Water 10(11):1513

IEA (2021) Kazakhstan energy profile. International Energy Agency (IEA), April 2020

IPCC (2014) Asia (Chapter 24). In: Climate change 2014: impacts, adaptation, and vulnerability. Contribution of Working Group II to the Fifth Assessment Report of the Intergovernmental Panel on Climate Change (IPCC)

IPCC (2022) Climate change 2022: impacts, adaptation and vulnerability. The Intergovernmental Panel on Climate Change (IPCC)

Janes C (2010) Failed development and vulnerability to climate change in Central Asia: implications for food security and health. Asia Pac J Public Health 22:236–245

Kerimray A, Kolyagin I, Suleimenov B (2018) Analysis of the energy intensity of Kazakhstan: from data compilation to decomposition analysis. Energ Effi 11(2):315–335

Kerimray A et al (2015) Climate change mitigation scenarios and policies and measures: the case of Kazakhstan. Clim Policy 16(3):332–352

Kim E, Standal K (2019) Empowered by electricity? The political economy of gender and energy in rural Naryn. Gend Technol Dev 23(1):1–18

Lioubimtseva E (2014) A multi-scale assessment of human vulnerability to climate change in the Aral Sea basin. Environ Earth Sci 73(2):719–729

Lioubimtseva E, Henebry GM (2009) Climate and environmental change in arid Central Asia: impacts, vulnerability, and adaptations. J Arid Environ 73(11):963–977

Lioubimtseva E, Cole R (2006) Uncertainties of climate change in arid environments of Central Asia. Rev Fish Sci 14(1–2):29–49

Lioubimtseva E et al (2005) Impacts of climate and land-cover changes in arid lands of Central Asia. J Arid Environ 62(2):285–308

Palazuelos E, Fernández R (2012) Kazakhstan: oil endowment and oil empowerment. Communis Post-Commun 45(1–2):27–37

Sabyrbekov R, Overland I (2020) Why choose to cycle in a low-income country? Sustainability 12(18):7775

Sabyrbekov R, Ukueva N (2019) Transitions from dirty to clean energy in low-income countries: insights from Kyrgyzstan. Cent Asian Surv 38. https://doi.org/10.1080/02634937.2019.1605976

Vakulchuk R (2016) Public administration reform and its implications for foreign petroleum companies in Kazakhstan. Int J Public Adm 39(14):1180–1194

Vakulchuk R, Overland I (2019) China's Belt and Road Initiative through the lens of Central Asia. In: Cheung FM, Hong Y-y (eds) Regional connection under the Belt and Road Initiative. The prospects for economic and financial cooperation. Routledge, London, pp 115–133

Vakulchuk R, Overland I (2021) Central Asia is a missing link in analyses of critical materials for the global clean energy transition. One Earth 4(12):1678–1692

Vakulchuk R, Daloz AS, Overland I, Sagbakken HF, Standal K (2022a) A void in Central Asia research: climate change. Cent Asian Surv 42:1–20

Vakulchuk R et al (2022b) Fossil fuels in Central Asia: trends and energy transition risks. Cent Asia Reg Data Rev 28:1–6

World Air Quality (2021) Air quality in Bishkek. https://www.iqair.com/kyrgyzstan/bishkek. Accessed on 16 Apr 2022

Rahat Sabyrbekov is an environmental economist who specializes in decarbonization, climate change and energy transition. Rahat is a Postdoctoral Fellow at OSCE Academy and Visiting Fellow at Davis Center at Harvard University. He obtained his PhD from School of Economics and Business at Norwegian University of Life Sciences. He received his Master's degree from University of Birmingham, the United Kingdom. He teaches Economics of Sustainable Management of Mineral Resources course at the OSCE Academy. His recent publications include *Putting the Foot Down: Accelerating EV Uptake in Kyrgyzstan* (2023), *Know Your Opponent: Which Countries might Fight the European Carbon Border Adjustment Mechanism?* (2022), *Fossil Fuels in Central Asia: Trends and Energy Transition Risks* (2022).

Indra Overland is Research Professor and Head of the Research Group on Climate and Energy at the Norwegian Institute of International Affairs (NUPI). He previously headed the Russia and Eurasia Research Group at NUPI and has worked on Central Asia since 2001. He completed his PhD at the University of Cambridge, followed by a three-year post-doctoral project on Central Asia and the South Caucasus. He has carried out fieldwork in all Central Asian states and has been responsible for cooperation between the OSCE Academy in Kyrgyzstan and NUPI since 2007. Every year he teaches MA students from all Central Asian countries on energy issues and hosts 4-5 Central Asian students in Norway. He is (co)author of *Caspian Energy Politics* (Routledge), *China's Belt and Road Initiative Through the Lens of Central Asia* (Routledge), *Kazakhstan: Civil Society and Natural Resource Policy in Kazakhstan* (Palgrave) and *Renewable Energy Policies of the Central Asian Countries* (CADGAT). He is a contributing author to the IPCC's Sixth Assessment Report (Working Group II).

Roman Vakulchuk is Senior Researcher at the Norwegian Institute of International Affairs (NUPI) in Oslo and holds a PhD degree in Economics from Jacobs University Bremen in Germany. He specializes in Central Asia and Southeast Asia and his main research interests are economic transition, trade, energy, climate change and investment policy. Vakulchuk has served as project leader in research projects organized by the Asian Development Bank (ADB), the

World Bank, the Global Development Network (GDN), the Natural Resource Governance Institute (NRGI) and others. In 2018, he worked as governance expert for OECD's mission in Kazakhstan and advised the government on privatization reform. In 2013, Vakulchuk was awarded the Gabriel Al-Salem International Award for Excellence in Consulting. Recent publications include *Seizing the Momentum. EU Green Energy Diplomacy Towards Kazakhstan* (2021), *Discovering Opportunities in the Pandemic? Four Economic Response Scenarios for Central Asia* (2020) and *Renewable Energy and Geopolitics: A Review* (2020).

Open Access This chapter is licensed under the terms of the Creative Commons Attribution 4.0 International License (http://creativecommons.org/licenses/by/4.0/), which permits use, sharing, adaptation, distribution and reproduction in any medium or format, as long as you give appropriate credit to the original author(s) and the source, provide a link to the Creative Commons license and indicate if changes were made.

The images or other third party material in this chapter are included in the chapter's Creative Commons license, unless indicated otherwise in a credit line to the material. If material is not included in the chapter's Creative Commons license and your intended use is not permitted by statutory regulation or exceeds the permitted use, you will need to obtain permission directly from the copyright holder.

Climate Research on Central Asia:
State of the Art

Climate Change: A Growing Threat for Central Asia

Anne Sophie Daloz ⓘ

Abstract Central Asia is highly vulnerable to climate change owing to a set of critical interactions between the region's socio-economic and environmental contexts. While some of the Central Asian countries are among the states contributing the least to global greenhouse gas emissions, they are already suffering directly from the effects of climate change. This chapter presents an overview of the physical impacts of climate change in Central Asia using the most recent literature, including the Sixth Assessment Report of the Intergovernmental Panel on Climate Change (IPCC). It identifies climate change-related risks and sectoral vulnerabilities for the region, providing background information to serve as context for the later chapters.

Keywords Central Asia · Climate change · Climate Impact Drivers (CID) · Sectoral vulnerability · Water scarcity

1 Central Asia's Climate

Central Asia is one of the most mountainous regions in the world, but has landscapes ranging from grasslands to deserts and woodlands in addition to its high mountains. The mountains of Central Asia consist of two major mountain ranges—the Pamir and the Tien Shan—which are critical to the livelihoods of the local communities. As shown in Fig. 1, the region is influenced by a variety of climates, with the desert and Mediterranean type environments in the south and continental climes everywhere else. The wide range of climates results in a high level of temporal and spatial variability in temperature and precipitation in the region. In most high-elevation areas, where the climate is dry and continental, there are hot summers and cool or cold winters with periodic snowfall. At lower elevations, the climate is mostly semi-arid to arid, with hot summers and mild winters with occasional rain and/or snow.

A. S. Daloz (✉)
CICERO, Oslo, Norway
e-mail: anne.sophie.daloz@cicero.oslo.no

© The Author(s) 2023
R. Sabyrbekov et al. (eds.), *Climate Change in Central Asia*,
SpringerBriefs in Climate Studies,
https://doi.org/10.1007/978-3-031-29831-8_2

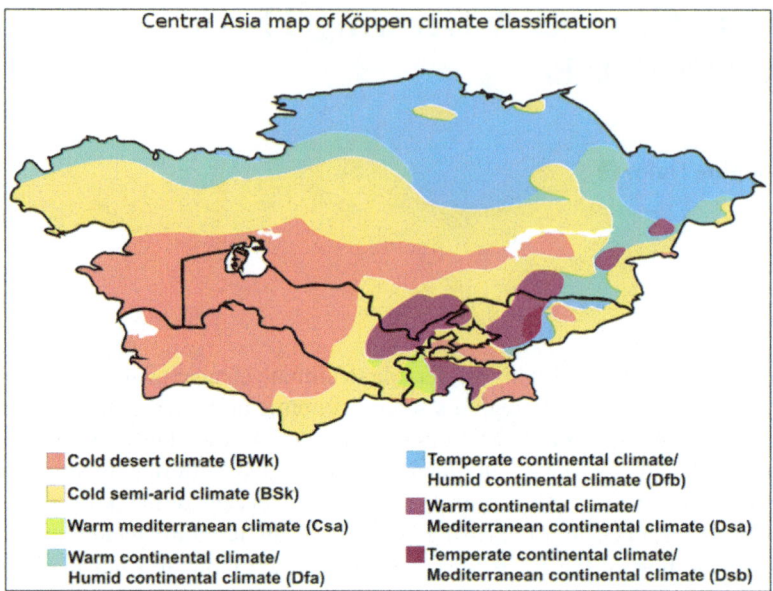

Fig. 1 Map of the different climates in Central Asia according to the Köppen climate classification. Credit: 'Central Asia map of Köppen climate classification'. *Source* World Köppen classification (with authors), enhanced, modified and vectorized by Ali Zifan. This figure is subject to the Creative Commons Attribution License (cc-by), license: CC-BY-SA-4.0, see https://creativecommons.org/licenses/by-sa/4.0/deed.en

2 Current and Projected Impacts of Climate Change

In 2021, Working Group I of the Intergovernmental Panel for Climate Change (IPCC) released its contribution to the Sixth Assessment Report in which they presented the most up-to-date physical understanding of the Earth's climate system and climate change. Once again, they pointed to the unequivocal human influence on our planet, and our warming of the atmosphere, oceans and land. This new report specifically examines the regional impacts of climate change—or Climate Impact Drivers (CIDs)—on the so-called 'climate reference regions'. These include 'West Central Asia' (WCA), which corresponds to our area of interest. For simplicity, we refer to the WCA region as 'Central Asia' in this chapter.

For Central Asia, the IPCC (2021) foresees several key changes in CIDs, including some that are predicted with 'high confidence'. The IPCC predicts a future increase in mean temperature and extreme heat as well as a decrease in cold spells and frost. They also report that, with the exception of a reduction in frost, these changes have already emerged in the historical period covered by their analysis. This is consistent with several studies (Hu et al. 2014; Zhang et al. 2019) showing how Central Asia has been warming faster than the global average over recent decades, with altitudinal variations (Haag et al. 2019) already affecting the region and local populations in multiple ways. These trends are concerning, particularly as the IPCC (2021) also predicts other changes with high confidence, for instance a loss of snow, glaciers and icesheets, declines in which have already been observed in several locations in the Pamir and Tien Shan mountain ranges (Sorg et al. 2012; Barandun et al. 2020). These changes have important implications in terms of water availability, a crucial and contentious resource in Central Asia (Vakulchuk et al. 2022). This issue could be aggravated further since the IPCC predicts—with medium confidence—an increase in aridity, alongside an increase in agricultural and ecological droughts. The combined increase in droughts and heatwaves can produce favorable conditions for wildfires, increasing the potential for burning across Central Asia (IPCC 2021) and the related risk of biodiversity loss (IPBES 2021).

Unlike mean temperature, changes in mean precipitation can be difficult to assess as climate models diverge on the direction and magnitude of the predicted changes (Christensen et al. 2013). The region's complex topography and lack of climate observations can at least partly explain the heterogeneous results. More recent work, however, is starting to shed light on expected future changes for Central Asia, such as a robust increase in annual mean precipitation, which is expected to be greatest over the Tian Shan mountains and the northern part of the region (Jiang et al. 2020). According to the 2021 IPCC report, the climate models also agree (with high confidence) on an increase in extreme precipitation leading to an increase in pluvial floods and associated landslides (the latter with medium confidence). Other types of flood risk are also expected: Zaginaev et al. (2019) have already observed an increase in glacial lake outburst floods over recent decades in Central Asia owing to its many high-altitude lakes and rapidly melting glaciers.

3 Sectoral Impacts and Vulnerabilities

Climate change has the potential to severely impact the livelihoods of local populations in Central Asia, with simultaneous and interlinked effects on the agriculture, energy, and transport sectors, as well as on public health. This section describes the ways in which the vulnerability of some of the most important sectors in Central Asia could increase in the future as a result of the effects of climate change.

3.1 Crop Production, Livestock and Food Security

In Central Asia, the majority of the population lives in rural areas and is highly dependent on agriculture and irrigation. The impacts of climate change on agriculture are diverse, and may be both positive and negative. Higher carbon dioxide concentrations and warmer temperatures can trigger an increase in crop yields (Orlov et al. 2021), whereas the impacts of extreme events (e.g. droughts, floods or heatwaves) can be devastating. The impacts on livestock should also not be omitted: climate change can affect both the quality of the feed and the health of the animals. The impacts on crops and livestock have obvious implications in terms of food security (see Standal et al., this volume).

3.2 Energy and Water Availability

The energy sector is also highly vulnerable to the effects of climate change as it is largely based on hydropower. Similarly to agriculture, the impacts on this sector can be both positive and negative. On the one hand, for some locations the increase in temperatures can lead to a diminution in the number of days where heating is required, therefore reducing the demand on the energy system. On the other hand, the current melting of glaciers and/or extreme precipitation can lead to floods, damaging infrastructure and leaving communities without electricity. In addition, the disappearance of glaciers, decreases in snow and/or changes to precipitation variability can also make water less available, increasing the pressure on the energy system as water is crucial for generating energy through hydropower. Furthermore, as more water will be needed for irrigation in the future, water use could become a source of tension between the energy and agriculture sectors. Water is already a contentious resource in Central Asia; a decrease in water availability could also exacerbate tensions between nations.

3.3 Health and Air Pollution

The expected increase in extreme temperatures will enhance heat stress for both urban and rural populations. In cities, more frequent and intense heatwaves can affect the health of the population, especially the most vulnerable, e.g. the youngest and the oldest (Meade et al. 2020). For rural populations, an increase in heat stress can affect the health of livestock, as well as impacting the health of the farmers and their ability to work outside (Orlov et al. 2021). Additionally, the impacts of climate change can lead to a decline in food quality via a decrease in nutrients, leading to increasing malnutrition (Myers et al. 2014). The potential increase in wildfires could also lead to an increase in air pollution, which is already a concern in urban areas in this region (UNDP 2021).

3.4 Transportation and Mobility

Transportation is interconnected with many sectors as it encompasses the mobility of people, energy and goods. In Central Asia, the projected effects of climate change (e.g. increase in flooding) can limit road access, limiting transportation or rendering it impossible. These impacts can have severe consequences for the livelihoods of local populations as well as the energy or agriculture sectors (see Blondin, this volume).

4 Conclusion

This chapter has provided an overview of the physical impacts of climate change in Central Asia and shown the increasing threat it represents for this region. According to the most recent literature, including the 2021 IPCC report, multiple changes in Climate Impact Drivers (CIDs) are expected in the future, including changes in mean temperatures and precipitation levels (rain and snow), as well as extreme events such as droughts, heatwaves and floods. The IPCC report also shows that some of these trends have already emerged in the historical period of their analysis. All these changes will increase the vulnerability of local populations via impacts to and across the energy, agriculture and transport sectors, as well as to public health (Vakulchuk et al. 2022). In this context, climate change can generate multiple tensions, including between sectors and between nations. For instance, the agriculture and energy sectors may find themselves fighting over water, an already contested resource in Central Asia, while questions of water scarcity and energy and food security could potentially cause or exacerbate tensions between Central Asian countries.

To better prepare for climate change and to limit its effects, mitigation and adaptation measures appropriate to the context of Central Asia are needed. To this end, more research is needed across the disciplines (Vakulchuk et al. 2022; Vakulchuk and Overland 2021) on, for example, extreme events and their multiple societal and sectoral impacts or on the linkages between gender and climate change. Some of this knowledge is presented in the following chapters of this book, but additional work is necessary to better assess the impacts of climate change in this region. This research will be key to designing the requisite adaptation and mitigation measures for Central Asia.

References

Barandun M, Fiddes J, Scherler M, Mathys T, Saks T, Petrakov D, Hoelzle M (2020) The state and future of the cryosphere in Central Asia. Water Secur 11:100072. ISSN 2468-3124. https://doi.org/10.1016/j.wasec.2020.100072

Christensen JH et al (2013) Climate phenomena and their relevance for future regional climate change. In: Climate change 2013: the physical science basis. Contribution of Working Group I to the Fifth Assessment Report of the Intergovernmental Panel on Climate Change. Cambridge University Press, Cambridge, pp 1268–1283

Climate Risk Profile (2021) Climate risk profile: Kyrgyz Republic. The World Bank Group and Asian Development Bank. https://www.adb.org/

Haag I, Jones PD, Samimi C (2019) Central Asia's changing climate how temperature and precipitation have changed across time, space, and altitude. Climate 7(10):123. https://doi.org/10.3390/cli7100123

Hu Z, Zhang C, Hu Q, Tian H (2014) Temperature changes in Central Asia from 1979 to 2011 based on multiple datasets. J Clim 27:1143–1167

IPBES (2021) Scientific outcome of the IPBES-IPCC workshop on biodiversity and climate change. https://www.ipbes.net/sites/default/files/2021-06/2021_IPCC-IPBES_scientific_outcome_20210612.pdf

IPCC (2021) Climate change 2021: the physical science basis (eds: Masson-Delmotte V, Zhai P, Pirani A, Connors S L, Péan C, Berger S, Caud N, Chen Y, Goldfarb L, Gomis MI, Huang M, Leitzell K, Lonnoy E, Matthews JBR, Maycock TK, Waterfield T, Yelekçi O, Yu R, Zhou B). Contribution of Working Group I to the Sixth Assessment Report of the Intergovernmental Panel on Climate Change. Cambridge University Press, Cambridge (in press)

Jiang J, Zhou T, Chen X, Zhang L (2020) Future changes in precipitation over Central Asia based on CMIP6 projections. Env Res Lett 15:054009

Meade RD, Akerman AP, Notley SR, McGinn R, Poirier P, Gosselin P, Kenny GP (2020) Physiological factors characterizing heat-vulnerable older adults: a narrative review. Environ Int 144:105909. ISSN 0160-4120. https://doi.org/10.1016/j.envint.2020.105909

Myers SS, Zanobetti A, Kloog I et al (2014) Increasing CO_2 threatens human nutrition. Nature 510(7503):139–142. https://doi.org/10.1038/nature13179

Orlov A, Daloz AS, Sillmann J et al (2021) Global economic responses to heat stress impacts on worker productivity in crop production. Econ Dis Cli Change 5:367–390. https://doi.org/10.1007/s41885-021-00091-6

Sorg A, Bolch T, Stoffel M et al (2012) Climate change impacts on glaciers and runoff in Tien Shan (Central Asia). Nature Clim Change 2:725–731. https://doi.org/10.1038/nclimate1592

UNDP (2021) Tackling air pollution in Europe and Central Asia for improved health and greener future. http://undp.org/

Vakulchuk R, Overland I (2021) Central Asia is a missing link in analyses of critical materials for the global clean energy transition. One Earth 4(12):1678–1692. https://doi.org/10.1016/j.oneear.2021.11.012

Vakulchuk R, Daloz AS, Overland I, Sagbakken HF, Standal K (2022) A void in Central Asia research: climate change. Cent Asian Surv: 1–20. https://doi.org/10.1080/02634937.2022.2059447

Zaginaev V, Petrakov D, Erokhin S, Meleshko A, Stoffel M, Ballesteros-Cánovas JA (2019) Geomorphic control on regional glacier lake outburst flood and debris flow activity over northern Tien Shan. Glob Planet Change 176:50–59. ISSN 0921-8181. https://doi.org/10.1016/j.gloplacha.2019.03.003; https://www.sciencedirect.com/science/article/pii/S0921818118306635

Zhang M, Chen Y, Shen Y, Li B (2019) Tracking climate change in Central Asia through temperature and precipitation extremes. J Geogr Sci 29:3–28

Anne Sophie Daloz is a climate scientist at CICERO, Norway. Her work relies on the analysis of climate data from observations to climate models outputs. She has worked on a variety of regions and processes such as Central Asia or Europe, and tropical cyclones, snowfall, clouds or precipitation. She also has a strong interest in connecting climate sciences to other disciplines.

Open Access This chapter is licensed under the terms of the Creative Commons Attribution 4.0 International License (http://creativecommons.org/licenses/by/4.0/), which permits use, sharing, adaptation, distribution and reproduction in any medium or format, as long as you give appropriate credit to the original author(s) and the source, provide a link to the Creative Commons license and indicate if changes were made.

The images or other third party material in this chapter are included in the chapter's Creative Commons license, unless indicated otherwise in a credit line to the material. If material is not included in the chapter's Creative Commons license and your intended use is not permitted by statutory regulation or exceeds the permitted use, you will need to obtain permission directly from the copyright holder.

Climate Change Science and Policy in Central Asia: Current Situation and Future Perspectives

Alisher Mirzabaev

Abstract Central Asia is already experiencing negative climate change impacts. Projections show that future climatic change will negatively affect many climate-sensitive economic activities in the region, particularly agricultural production and associated livelihoods. Mitigating and adapting to climate change in Central Asia requires a significant increase in investment in climate change research, as well as the mainstreaming of adaptation actions into public policies. This paper assesses the current state of climate change science in the region and the key trends, based on a bibliometric and content analysis review. It provides a perspective on investment priorities for climate change-related research, as well as measures that will build synergies between climate actions and other priorities for sustainable development in the region. The paper calls for an expansion in open access to data; increased investment in climate change research, especially in the social sciences; development of local climate change modelling capacities; and support for regional knowledge and scientific exchange on the topic of climate change.

Keywords Bibliometric analysis · Climate change literature · Science policy · Research investment · Central Asia

1 Highlights

- The climate crisis is already having a strong impact on many aspects of the social and economic lives of Central Asia populations, and these are set to increase.
- The scientific literature on climate change in Central Asia has also been growing rapidly, but remains very small.
- Promising areas for investment to promote climate change science in Central Asia include improving open access to data, investing more in the social sciences,

---

A. Mirzabaev (✉)
Institute of Food and Resource Economics (ILR), University of Bonn, Bonn, Germany
e-mail: almir@uni-bonn.de

developing local climate modelling capacities and supporting the emergence of regional scientific journals focusing on climate change and broader environmental issues.

2 Introduction

The climate crisis is already having a significant impact on many aspects of people's social and economic lives in all regions of the world (IPCC 2021, 2022), while also influencing international relations in terms of trade, political and economic alliances and scientific collaboration (Friel et al. 2020; Mirzabaev et al. 2021; Ortiz et al. 2021). Central Asia is deeply involved in many of these processes and significantly affected by climate change, in some ways more so than other regions of the world. In this context, science plays a critically important role in enhancing people's understanding of the present and future impacts of climatic changes, and in climate change adaptation and mitigation in the region. Over the last two decades, globally the scientific literature on climate change has grown exponentially (Wheeler and von Braun 2013; Nalau and Verrall 2021; IPCC 2022). The climate change theme is also advancing, slowly but surely, to the forefront of policy discussions in Central Asian countries. Although climate change does not dominate national conversations in the region in the same way as it does in some other countries around the world, regional decision makers are increasingly attentive to issues related to climate change impacts, loss and damage and climate change adaptation and mitigation, because global efforts to address climate change are reshaping the nature of international relations, international trade and global economic competitiveness. Moreover, observed changes in extreme weather events (e.g. heatwaves in summer, more frequent dust storms, droughts) are starting to influence social awareness of climate change in the region and peoples' expectations from their governments to deal with this problem.

In parallel with these evolving policy and social contexts in Central Asia, the scientific literature on climate change in Central Asia has also been expanding rapidly. However, as we will see in the following sections, the current state of research activities and scientific outputs from Central Asian countries on climate change issues remain insufficient. To ensure successful development in Central Asia, science needs to provide viable policy advice, information, knowledge and solutions for the climate crisis. The objective of this paper is to assess the current state of climate change science in the region, identify major areas of success and key gaps and provide a perspective on investment priorities for climate change-related research and development. This paper, thus, adds to the emerging literature on the state and priorities for climate change research in Central Asia (Vakulchuk et al. 2022). This paper seeks to help fill this critical gap, highlighting promising directions for climate change science investment in the region.

3 Methodology

A number of methodologies are frequently used to analyse the evolution and contemporary state of scientific literature. These include systematic reviews, meta-analyses, narrative reviews (Greenhalgh et al. 2018; Harari et al. 2020; Rethlefsen et al. 2021) and bibliometric, visualisation and content analysis reviews (Nalau and Verrall 2021). In this paper, a bibliometric and content analysis approach is used, because this approach involves analysis of the meta-data of publications on climate change in Central Asia in a quantitative way that enables us to summarise large amounts of very diverse literature according to specific dimensions, such as key themes, key authors and major donors supporting the research behind the publications. Systematic reviews, meta-analyses and narrative reviews are usually used to answer very specific theme-focused research questions, and cannot provide an easy and visually accessible overview of such diverse literature spanning multiple disciplines.

The underlying databases for this analysis come from the Scopus and Web of Science indexing services. In each of these indexing services, a literature search was conducted using key words: 'climate change', in combination with 'Central Asia', 'Kazakhstan', 'Kyrgyzstan', 'Tajikistan', 'Turkmenistan' and 'Uzbekistan' in the titles of the publications. The search was done on the publication titles to ensure that the selected papers were specifically focused on climate change issues in Central Asia or in specific Central Asian countries, rather than focusing only briefly on Central Asia while having a broader geographic coverage.

In addition, a similar search was conducted in Google scholar and the Russian Language Scientific Electronic Library eLIBRARY.RU, in order to triangulate the findings from the Scopus and Web of Science indexing services. Scopus and Web of Science are primarily focused on literature published in the English language in scientific journals and books, whereas Google Scholar also contains information about other forms of publication, such as doctoral theses, preprints and technical reports. The scientific community in Central Asia actively uses the Russian language in research and writing, hence the use of eLIBRARY.RU, which indexes scholarly work in Russian.

The highest return of publications was in the Google scholar search (601), while the eLIBRARY.RU search resulted in 143 publications, Scopus—185 and Web of Science—166. All of the indexes have been used (in different ways) in the analysis that follows. However, since Scopus and Web of Science are expected to index high quality scientific literature, which has been exposed to rigorous international peer review, the content analysis has been limited to the papers indexed in those sources. Moreover, there is a significant overlap between publications indexed in Scopus, Web of Science and Google Scholar, but only little overlap with those publications written in Russian. For this reason, although the analysis presented here is considered to be largely representative of the key study themes and disciplines, it is not fully representative of the sources of funding. It was not possible to collect information about funding sources from the Russian language publications indexed in eLIBRARY.RU. Publications produced in Russian are more likely to be funded by local or Russian sources.

4 Current State of Climate Change Science in Central Asia

The search results in Google Scholar provide a comprehensive overview of English-language publications on the topic of climate change in Central Asia (Fig. 1). These results highlight a rapid growth in the scientific literature on climate change issues in the region over the last two decades, but especially during the last five years. Most publications (325) have a regional character, i.e. they investigate aspects of climate change relevant for the entire Central Asian region. In addition, there are country-specific publications on climate change, with the largest number focusing on Kazakhstan (104), followed by Uzbekistan (69), Tajikistan (65), Kyrgyzstan (35) and Turkmenistan (5). For comparison, a similar literature search in Google Scholar for China returned 6770 publications, for Pakistan—921 and for Mongolia—407 publications, all of which are considerably higher than for any individual country in Central Asia, even after taking into account the differences in populations. While the total numbers of publications, both at regional level and for specific Central Asian countries, are therefore highly insufficient. The very few publications dedicated to climate change issues in Turkmenistan are of particular concern.

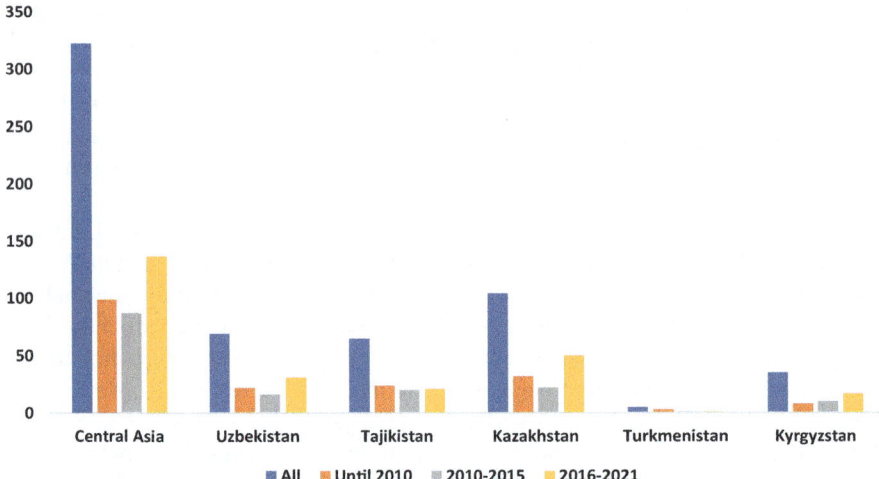

Fig. 1 The number of publications specifically devoted to climate change issues in Central Asia (in English language). *Source* Based on Google Scholar search (as of 10.12.2021)

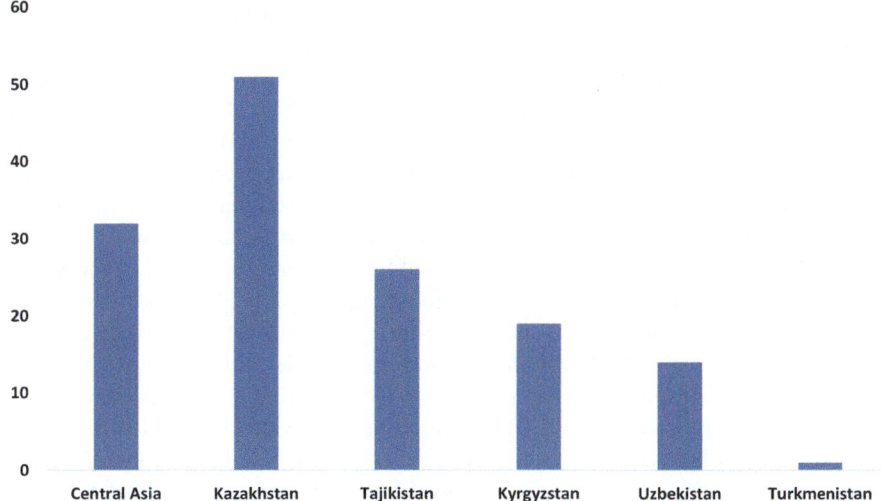

Fig. 2 The number of publications specifically devoted to climate change issues in Central Asia in Russian language. *Source* Based on Scientific Electronic Library eLIBRARY.RU (as of 22.02.2022)

These numbers do not change significantly when Russian language publications are accounted for (Fig. 2). Here, eLIBRARY.RU indexes 32 publications for Central Asia as a whole. By country, the figures were 51 publications for Kazakhstan, followed by Tajikistan (26), Kyrgyzstan (19), Uzbekistan (14) and Turkmenistan (1). This broadly confirms the patterns in the Google Scholar search, although publications on climate change focusing on Uzbekistan appear more likely to have been written in English rather than Russian, while significantly fewer have been written about Central Asia as a whole in Russian compared to the English-language publications. There were also significantly fewer publications written about Central Asia as a whole in Russian compared to English-language publications (32 compared to 325).

The analysis of the content of the abstracts of the peer-reviewed journal papers, books and book chapters indexed by Scopus shows that they could be broadly clustered into five themes: (1) ecosystem services, biodiversity and the carbon cycle; (2) paleoclimate; (3) water resources; (4) crop production; and (5) environmental management and land-use change. No strict allocation of publications into these five clusters is possible, since any given publication may have contributed to two or more of these clusters.

These results show that publications on climate change in Central Asia have been primarily dedicated to the biophysical impacts of climate change. There have been relatively few studies investigating the socioeconomic impacts of climate change, issues related to people's vulnerability to climate change and the social and economic dimensions of climate change adaptation. This is indicated by the failure of related words ('socioeconomic', 'human vulnerability', 'social impacts', 'adaptation', etc.) to emerge prominently during the textual analysis of these publications.

These results are corroborated by disciplinary classification of the climate change literature on Central Asia indexed by Web of Science, which also clearly highlights the under-representation of social science publications compared to other disciplines. Only about 5% of those publications cover socioeconomic topics, while 95% of the publications focus on various natural sciences, predominantly, environmental sciences and ecology, geosciences, physical geography and water resources.

Analysis of the authors who wrote these publications indicates that a majority of these peer-reviewed publications on climate change impacts in Central Asia have been authored by Chinese researchers. The analysis of linkages between authors reveals a dominant role of Chinese researchers and smaller number of publications by other research groups made up of authors from Central Asia and other countries, mainly Europe and North America. Moreover, there are close and strong links between various Chinese researchers through joint authorship of publications. In this regard, a particularly important role is played by the Xinjiang Institute of Ecology and Geography of the Chinese Academy of Sciences in Urumqi, which practically serves as the central hub of Chinese research on climate change issues in Central Asia. There are some peripheral co-authorship collaborations between Chinese authors and Central Asian authors, but overall it appears that this is relatively limited with all-Chinese author groups dominating most of these publications.

The peer-reviewed English-language journals which have most frequently been published on climate change in Central Asia are The Science of the Total Environment and Quaternary International. Other frequently used journals are Sustainability, the Journal of Arid Land, Global and Planetary Change, and Sustainability. In this regard, it is worthwhile to note that the Journal of Arid Land is published by Springer in collaboration with the Xinjiang Institute of Ecology and Geography.

Overall, the results show that the number of scientific publications on climate change in Central Asia has been growing rapidly, but remains relatively low compared to publications on climate change relating to other neighbouring world regions. A majority of these publications was produced by Chinese researchers and other international research groups. Only a relatively limited number of these publications were written exclusively by local research teams. Almost all of these publications were written under natural science disciplines, with extremely few of the publications coming from the social sciences.

Socio-economic research on climate change impacts, evolving vulnerabilities to climate change, distributional dimensions of climate change impacts and analysis of adaptation options and policy responses to climate change represents an absolutely essential enabling element for successful adaptation to climate change. The current limited nature of such research in the region is a very grave threat to the sustainable development of Central Asian countries in the context of a changing climate, even in the short- to medium-term in the next 2–3 decades. Central Asian countries might be intending to deal with the emerging challenges of climate change as they emerge. However, ad hoc measures in response to emerging weather and climatic extremes are both costly and economically inefficient (Gerber and Mirzabaev 2017). What is needed is proactive actions to strengthen adaptive capacities and reduce vulnerabilities to climate change in the region. Increasing investment into climate change research, particularly in the social sciences, would be a crucial cornerstone of such a transition towards smart, proactive and efficient climate change adaptation policies.

5 Key Areas for Investment in Climate Change Science in Central Asia

The textual analysis of information about the donors who funded the publications on climate change in Central Asia clearly highlights the dominance of Chinese government organisations, including the Chinese Academy of Sciences and National Science Foundation of China. The analysis also indicates the involvement of European science funders such as German Research Foundation (DFG), the German Federal Ministry of Education and Research (BMBF) and the United Kingdom's UK Research and Innovation (UKRI). The government of Kazakhstan is also represented among these donors, but the national science donors from the other countries of Central Asia are not prominent among the donors of publications in reputed international journals. This can be partially explained by the linguistic dimension, with many locally funded climate change-related publications being primarily published in local languages or Russian. However, even after some level of accounting for this, the overall picture in terms of local funding for climate change research comes out to be very insufficient. This calls for a rapid and substantial increase in investment from the Central Asian governments for climate change research.

In this regard, in my view, the major opportunities for research investment on climate change science that will create multiple beneficial synergies between climate change adaptation and other priorities for sustainable development in the region are the following:

Open access to data. For expanding research on climate change, the meteorological and statistical agencies across Central Asian countries need to provide easy and open access to long-term weather and hydrological data, high-resolution statistics on agricultural, environmental and land-use impacts (e.g. at district levels), as well as to the results of agricultural and other household surveys, as is done in other countries and regions in the world. Currently, there is a dearth of household survey results available for Central Asian countries. More investment needs to be directed towards conducting representative and periodic surveys of households and of climate change impacts on households. Naturally, data privacy issues need to be addressed before giving public access to the results of any such household surveys, but these are technical issues that can easily be resolved. Without access to these sources of data, there is very little opportunity for rigorous socioeconomic research on climate change in the region.

More investment in climate change research in the social sciences. This is particularly important for research on such themes as climate change impacts, vulnerability and adaptation. It will be extremely inadequate for decision making on climate change adaption policies if the available research is only on the environmental dimensions of climate change. Policymakers' major interest in climate change research is due to their need to understand better how climate change affects people and societies. Environmental research on climate change provides a valuable first entry point in understanding climate change, but it is essential that social sciences then use that environmental research to investigate climate change impacts on people and propose solutions. Without such social science research happening in parallel, environmental research on climate change by itself has a relatively limited social value.

Developing local climate modelling capacities. There is still a huge gap in terms of high-resolution local projections of climate change in the region. Global models, whose projections are often used in describing future climate changes in the region, do not fully take into account a myriad of local factors in sufficient detail. These local factors will have a crucial effect on the way in which regional climate change impacts (e.g. local land-use changes, sand and dust storms, the role of irrigation, the knock-on effects of the Aral Sea desiccation, the role of water reservoirs). Hence, investment needs to be directed to develop highly localised models of climate change forecasting.

Establish regional journals focusing on climate change. There is a strong need to boost scientific collaboration, exchange of knowledge and information on climate change among the research communities in Central Asian countries. Despite the rapid spread of English language use in scientific publications in the region over the last decade, still the lack of knowledge of English remains an important barrier to international publishing of climate change research produced by Central Asian researchers. As a result, many studies conducted in the region do not benefit from high quality, rigorous and constructive peer-review processes. The establishment of regional scientific journals dedicated to climate change, adopting transparent, high-quality peer-review procedures (e.g. following the peer-review style of the Frontiers journals), will help both to improve the quality of scientific publications and to promote mutual learning.

6 Conclusions

Climate change projections show that Central Asia will experience major climatic changes in the coming decades. This requires well-planned proactive measures to adapt to these changes. Climate change research should play an essential role in this adaptation process. However, the current state of climate change science in the region is not up to the task. There is a need for a significant increase in investment in climate research in all countries of Central Asia. Potential priority areas with a high social return on investment might include: expanding open access to data; increasing investment in climate change research (especially in the social sciences); developing local climate change modelling capacities; and facilitating regional knowledge and scientific exchanges on climate change, for instance through the establishment of regional scientific journals.

References

Friel S, Schram A, Townsend B (2020) The nexus between international trade, food systems, malnutrition and climate change. Nature Food 1(1):51–58

Gerber N, Mirzabaev, A (2017) Benefits of action and costs of inaction: drought mitigation and preparedness—a literature review. 1. WMO, Geneva, Switzerland and GWP, Stockholm, Sweden

Greenhalgh T, Thorne S, Malterud K (2018) Time to challenge the spurious hierarchy of systematic over narrative reviews? Eur J Clin Investig 48(6):e12931

Harari MB et al (2020) Literature searches in systematic reviews and meta-analyses: a review, evaluation, and recommendations. J Vocat Behav 118:103377

IPCC (2021) Climate change 2021: the physical science basis (eds: Masson-Delmotte V, Zhai P, Pirani A, Connors SL, Péan C, Berger S, Caud N, Chen Y, Goldfarb L, Gomis MI, Huang M, Leitzell K, Lonnoy E, Matthews JBR, Maycock TK, Waterfield T, Yelekçi O, Yu R, Zhou B). Contribution of Working Group I to the Sixth Assessment Report of the Intergovernmental Panel on Climate Change. Cambridge University Press, Cambridge, UK and New York, NY, In press, doi:https://doi.org/10.1017/9781009157896.

IPCC (2022) Climate change 2022: impacts, adaptation, and vulnerability (eds: Pörtner H-O, Roberts DC, Tignor M, Poloczanska ES, Mintenbeck K, Alegría A, Craig M, Langsdorf S, Löschke S, Möller V, Okem A, Rama B). Contribution of Working Group II to the Sixth Assessment Report of the Intergovernmental Panel on Climate Change. Cambridge University Press. Cambridge University Press, Cambridge, UK and New York, NY, 3056 pp. https://doi.org/10.1017/9781009325844

Mirzabaev A et al (2021) Climate change and food systems. Center for Development Research (ZEF) in cooperation with the Scientific Group for the UN Food System Summit 2021. https://bonndoc.ulb.uni-bonn.de/xmlui/handle/20.500.11811/9206

Nalau J, Verrall B (2021) Mapping the evolution and current trends in climate change adaptation science. Clim Risk Manag 32:100290. https://doi.org/10.1016/j.crm.2021.100290

Ortiz AMD et al (2021) A review of the interactions between biodiversity, agriculture, climate change, and international trade: research and policy priorities. One Earth 4(1):88–101

Rethlefsen ML et al (2021) PRISMA-S: an extension to the PRISMA statement for reporting literature searches in systematic reviews. Syst Rev 10(1):1–19

Vakulchuk R, Daloz AS, Overland I, Sagbakken HF, Standal K (2022) A void in Central Asia research: climate change. Cent Asian Sur. https://doi.org/10.1080/02634937.2022.2059447

Wheeler T, von Braun J (2013) Climate change impacts on global food security. Science 341:508–513. https://doi.org/10.1126/science.1239402

Alisher Mirzabaev is the Interim Chair of the Production Economics Group, Institute of Food and Resource Economics (ILR) at the University of Bonn. Before joining ILR, he was a senior researcher at the Center for Development Research (ZEF), University of Bonn. Previously, he also worked as an economist with the International Center for Agricultural Research in the Dry Areas (ICARDA). Dr. Mirzabaev was a lead researcher in several international projects with both global and regional focus on Africa and Asia. He was a Coordinating Lead Author of the Chapter on Desertification of the IPCC Special Report on Climate Change and Land. He has a PhD degree in agricultural economics from the University of Bonn, Germany. He has authored more than 100 scientific publications, including peer-reviewed journal articles, books and book chapters, and discussion papers. His research interests include climate change impacts and adaptation, the economics of land degradation and ecosystem restoration, and the water–energy–food security nexus.

Open Access This chapter is licensed under the terms of the Creative Commons Attribution 4.0 International License (http://creativecommons.org/licenses/by/4.0/), which permits use, sharing, adaptation, distribution and reproduction in any medium or format, as long as you give appropriate credit to the original author(s) and the source, provide a link to the Creative Commons license and indicate if changes were made.

The images or other third party material in this chapter are included in the chapter's Creative Commons license, unless indicated otherwise in a credit line to the material. If material is not included in the chapter's Creative Commons license and your intended use is not permitted by statutory regulation or exceeds the permitted use, you will need to obtain permission directly from the copyright holder.

Central Asian Decarbonisation Pathways and Carbon Pricing

Central Asian Climate Policy Pledges Under the Paris Agreement: Can They Be Fulfilled?

Rahat Sabyrbekov, Indra Overland, and Roman Vakulchuk

Abstract The Central Asian region has been and will continue to be significantly impacted by climate change and all the region's countries have pledged nationally determined contributions (NDCs) under the Paris agreement. This chapter aims to assess how likely Central Asian countries are to fulfil these pledges. To answer this question, we compare the NDCs to their respective national development programmes and historical trends. The results show that the countries of Central Asia vary in their ability to fulfil their pledges and that doing so will require structural changes to their energy systems, substantial investments in infrastructure and, most importantly, the alignment of their development plans with their declared climate goals. None of the countries have thus far engaged in structural reforms aimed at large-scale climate change adaptation and mitigation.

Keywords Central Asia · Climate policy · Development · Energy · Energy transition · Green economy

1 Introduction

The impact of climate change in Central Asia will be devastating. The existing research predicts increased risk of droughts and other natural hazards, rising prevalence of airborne diseases, the disappearance of many glaciers and the occurrence of unprecedented heatwaves (IPCC 2022). The research further indicates that the most vulnerable sectors of the economy are energy, agriculture, infrastructure and

R. Sabyrbekov (✉)
Organization for Security and Cooperation in Europe (OSCE) Academy, Bishkek, Kyrgyzstan
e-mail: r.sabyrbekov@osce-academy.net

I. Overland · R. Vakulchuk
Norwegian Institute of International Affairs (NUPI), Oslo, Norway
e-mail: ino@nupi.no

R. Vakulchuk
e-mail: rva@nupi.no

© The Author(s) 2023
R. Sabyrbekov et al. (eds.), *Climate Change in Central Asia*,
SpringerBriefs in Climate Studies,
https://doi.org/10.1007/978-3-031-29831-8_4

healthcare (Reyer et al. 2017). The countries of the Central Asian region do not have the resources needed to combat the negative consequences of climate change (Vakulchuk et al. 2022). Therefore, the region urgently needs to adopt sound national policies to adapt to and mitigate the risks.

Under the Paris Climate Agreement, the Central Asian governments have submitted nationally determined contributions (NDCs) and developed national adaptation plans (NAPs). It remains to be seen how realistic these goals and policies are in terms of their implementation. The countries have also announced climate strategies, both as part of their international climate commitments and within their national development plans. Inclusion of the impacts of climate change and strategies to mitigate or adapt to these in the national development plans of Central Asian countries is particularly important because of the region's climate vulnerability.

Despite this, we found no literature assessing the stated NDCs of the Central Asian countries and their alignment with national development strategies. This leaves key questions unanswered with regard to the content of the pledges, whether the countries are able to fulfil them based on current and historical data, and how well the stated national development goals align with each country's climate goals.

This chapter examines whether four Central Asian countries—Kazakhstan, Kyrgyzstan, Tajikistan and Uzbekistan—can achieve their climate policy goals and whether their national plans match these goals. The fifth Central Asian country, Turkmenistan, is excluded from the study as no historical data or information on their strategy is available. To answer the questions posed above, we briefly review the historical trends in the national energy sectors and emissions. We then compare each country's NDC with its current national development strategy. The NDCs were accessed from the UNFCCC NDC registry in March 2022. The latest national development programmes were retrieved from the official websites of the governments of each country at the same time.

2 Historical Trends

2.1 Energy: Growing Reliance on Fossil Fuels

Central Asia is rich in fossil fuels and their importance for the economies of the region is hard to overestimate (Vakulchuk et al. 2022). The energy supply sector has a heavy historical legacy and has not changed much in the last fifty years. Thus, the energy infrastructure is quickly becoming outdated and lacks investment. The hydropower plants in Kyrgyzstan and Tajikistan, for example, are crumbling and require significant updating (Bekchanov et al. 2015; World Bank 2017). Historically, Kazakhstan and Uzbekistan have relied on oil and gas, while Kyrgyzstan and Tajikistan have been dependent on hydropower in addition to fossil fuels.

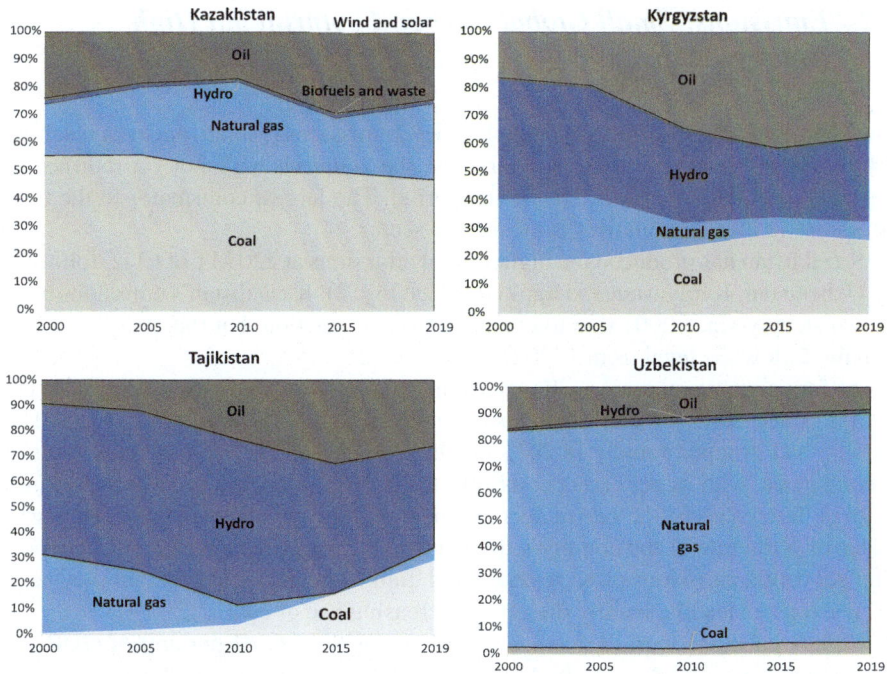

Fig. 1 Total primary energy supply by source (in per cent). *Source* IEA (2022)

The composition of the energy supply of the countries from 2000 to 2020 remained stable, with fossil fuel use making up the largest share (see Fig. 1). The energy mix in the total energy supply of Kazakhstan has not changed significantly, with hydrocarbons providing about 70% of all energy and only a small fraction coming from renewable energy in recent years.

In the cases of Kyrgyzstan and Tajikistan, we can observe the growing importance of coal and oil against the backdrop of a declining share of hydropower. This is due to the inability of the existing hydro plants to meet the growing energy demand from the residential and industrial sectors. From 2010 onwards, the Tajik government increased its reliance on coal for both electricity and heat. A similar pattern is found in Kyrgyzstan where it is coal and oil consumption that have grown (Vakulchuk et al. 2022).

Outdated energy infrastructure increases the vulnerability of the Central Asian countries to climate change. Most of the current energy infrastructure in Central Asia was built in the 1970s and has seen a little modernization since (Laldjebaev et al. 2018; Sabyrbekov and Ukueva 2019). The cost of modernization is estimated to be significant, and the region has not managed to attract the foreign direct investment needed to upgrade the infrastructure (Vakulchuk 2022).

2.2 Emissions: Small Global Carbon Footprint but High Energy Intensity

The contribution of Central Asia to global GHG emissions is small, not even reaching 1% of global emissions in 2018. Moreover, the countries have not yet returned to the level of emissions during the Soviet period. The largest contributor to the total emissions in all four countries is the energy sector.

Kazakhstan has produced the highest total emissions at 220 MT of CO_2, followed by Uzbekistan, Kyrgyzstan and Tajikistan (see Fig. 2). Kazakhstan's emissions have grown steeply since 2001 due to expanding oil production, but the 2020 level still remains below the level from 1990 (Fig. 3).

As for per capita emissions, Kazakhstan also produces the highest amount in the region. Since 2009, its emissions have stabilized at 12–15 tonnes per capita. The amount had dropped rapidly in 2013 and then stagnated. Uzbekistan finds itself in second place, with its per capita emissions falling from 6 tonnes in 1990 to 4 tonnes in 2018. Interestingly, the rate of decrease in emissions per capita in Uzbekistan was faster than the rate of the country's total emissions, suggesting a gain in efficiency. The per capita emissions of Kyrgyzstan and Tajikistan are the lowest, but they have been growing steadily, mainly driven by increasing use of coal.

Compared to the world's average energy intensity of 6 MJ per unit of GDP, the economies of Central Asia are highly energy intensive (see Fig. 4). The highest level of energy intensity is observed in Uzbekistan, whereas Tajikistan has the lowest level. The energy intensity of the Tajik economy has been decreasing steadily and, since 2009, has been below the world average. In Kazakhstan, the level fell in 2001 and has remained stable since then. Though it has the highest level, Uzbekistan has also substantially reduced the energy intensity of its economy since 2003. From 2001

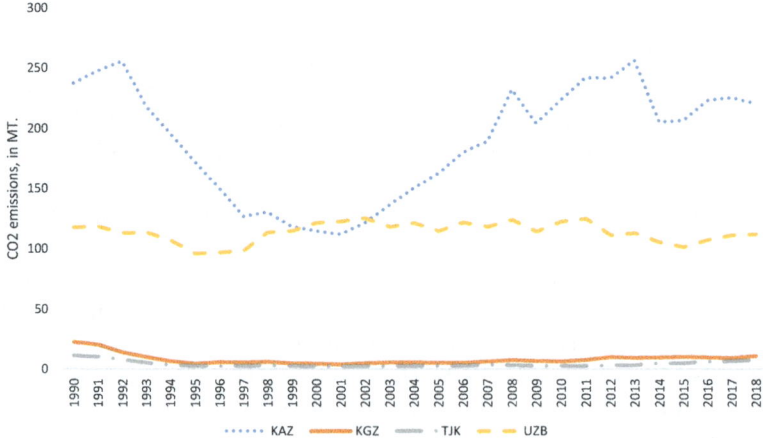

Fig. 2 Total CO_2 emissions by country (in MT). *Source* World Bank (2022)

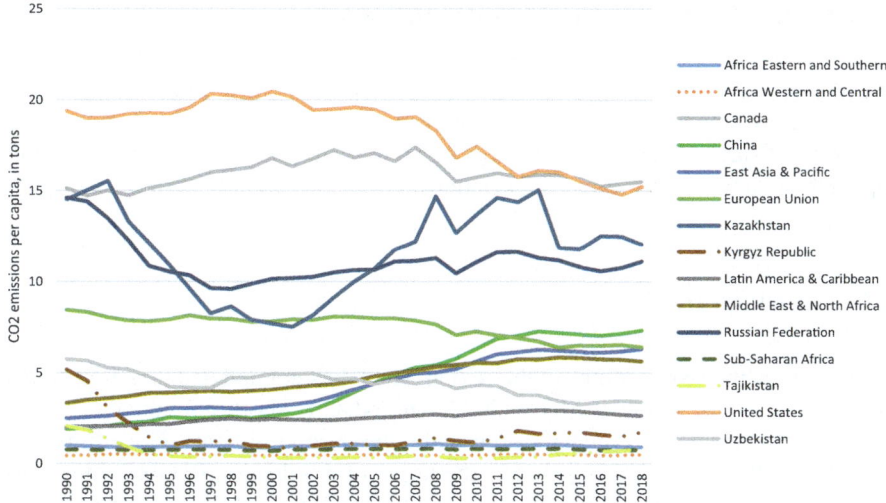

Fig. 3 CO_2 emissions per capita of Central Asian countries and selected world regions and countries

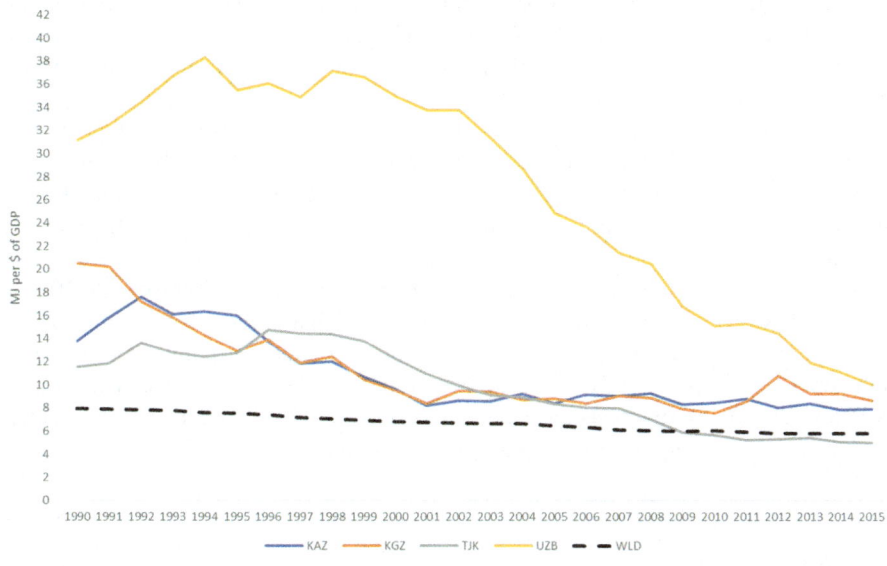

Fig. 4 Energy intensity (MJ/$ 2011 PPP GDP). *Source* World Bank (2022)

onwards, Kyrgyzstan's energy intensity grew fast, becoming second highest in the region in 2015. The high energy intensity of the Central Asian economies is partially explained by outdated infrastructure, lack of energy efficiency regulations and limited investment.

3 Nationally Determined Contributions

The four countries covered in this chapter are signatories to the Paris Climate Agreement and have submitted national development contributions (NDCs) that envisage the taking of various steps to reduce GHG emissions. The level of detail and overall approach of each country varies substantially.

3.1 Kazakhstan

Kazakhstan is the only country in the region which has not updated its initial NDC, submitted in 2016, as of 19 March 2022. The NDC is short and contains only a few details regarding the country's proposed climate commitments (Government of Kazakhstan 2016). The document sets the target of a reduction in GHG emissions by 15% by 2030 based on Kazakhstan's own resources, and 25% if international support is provided. The base year is 1990.

The country aims to reduce emissions in all sectors of its economy. The priority sectors are energy, agriculture, waste, land use and forestry. The NDC emphasizes increased energy efficiency and deployment of renewable energy. It also envisages improvements in waste management, modernization of housing and promotion of sustainable transport.

Despite the non-updated NDC, Kazakhstan is the regional climate policy leader. The country was the first and only state in Central Asia thus far to establish a national Emissions Trading Scheme, which it did in 2013 (Abdi et al., this volume). Kazakhstan was also the first in the region to attract sizable financial support from international banks for its renewable energy projects. In 2020, the share of renewables in the total energy supply of Kazakhstan reached a measly 3% (Vakulchuk et al. 2022).

In 2020, Kazakhstan's President Tokayev declared that the country would achieve carbon neutrality by 2060. Despite this ambitious declaration, many doubt that the high dependence on fossil fuels can be overcome so soon (e.g. Poberezhskaya and Bychkova 2021). Internal resistance and external geopolitical challenges may greatly complicate the ongoing decarbonization of the country (Koch and Tynkkynen 2021).

3.2 Kyrgyzstan

In 2021, Kyrgyzstan submitted an updated NDC which aims to reduce GHG emissions by 44% by 2030 with international support and by 16% without international support (Government of Kyrgyzstan 2021b). A business-as-usual scenario was used as baseline.

According to its NDC, Kyrgyzstan seeks to implement reforms in the energy, agriculture, forestry and land use sectors. In the energy sector, the country aims to increase the share of renewable energy, improve energy efficiency, improve the natural gas supply network and increase the number of electric vehicles.

In the agricultural sector, the main activities include improvements in livestock productivity, increased organic crop production, increased efficiency of manure use and the generation of biogas. In forestry and land use, the measures include an increase in forest coverage and expansion of perennial plantations. These measures should contribute to both climate change adaptation and mitigation.

In Kyrgyzstan, 93% of all GHG reductions are to be achieved in the energy sector. The reductions are mainly (37%) to be achieved through a shift from the burning of coal to natural gas for heating, to be facilitated by the gasification of households (Table 1).

Table 1 Planned activities to reduce GHG emissions in the Kyrgyz energy sector under the most optimistic scenario (% of total reduction)

Activity	Measure	Share in total GHG emission reduction (%)
GHG reduction	Gasification of households	16
	Electric private vehicles	12
	Improved traffic management and cycling infrastructure	7
	Reduction in electricity loss in transmission and distribution	1
	Gasification of public transport	0.4
	Heat supply modernization	0.1
	Public transport electrification	>0.01
Energy efficiency improvement	Energy efficient stoves for households	20
	Replacement of coal-fired boilers with gas	15
	Energy efficiency improvements to buildings	0.5
Renewable energy deployment	Biogas expansion	22
	Construction of new small hydro plants	2
	Geothermal energy	2
	Expansion of solar heat collectors	1
	Solar power development	0.6
	Wind energy development	0.2
	Modernization of existing heat power plants	>0.1
	Small private hydro	>0.1

Source Authors' calculations based on the NDC of Kyrgyzstan

The Kyrgyz NDC includes an adaptation section with measures to address the most vulnerable sectors such as water resources, agriculture, energy infrastructure, public health, and forestry and biodiversity. It also features new intersectoral sections on Climate-Resilient Areas and Green Cities.

The Kyrgyz NDC may look ambitious in terms of overall reduction of GHG emissions, but the actual implementation relies heavily on the expansion of the gas supply network, i.e. an increase in fossil fuel use. Other significant reductions depend on external financial and technical support, for example for biogas expansion and the modernization of household heating. The role of renewable energy is small and the reduction of GHG emissions due to wind and solar power is expected to contribute to only about 1% of the total GHG reduction in the energy sector. Moreover, the government currently has no plans for a large-scale transition to clean energy.

3.3 Tajikistan

Tajikistan submitted an updated version of its NDC in 2021 (Government of Tajikistan 2021). The target for GHG emissions is to not exceed 60–70% of the level of 1990 emissions. With international support, the goal is to not exceed 50–60% of this level.

The priority sectors include agriculture, energy, forestry and biodiversity, industry and construction, transport and infrastructure. However, Tajikistan has neither provided sector-specific GHG emission targets, nor stated the concrete actions it will take. The measures are broad and mention the promotion of renewable energy, increased resilience on the existing energy and transport infrastructure, improvement of energy efficiency and promotion of non-motorized transport.

Ultimately, the NDC focuses strongly on adaptation and increasing resilience to climate change impacts. Specifically, the document emphasizes the strategic importance of the modernization of the existing energy infrastructure such as hydropower plants.

3.4 Uzbekistan

The updated NDC of Uzbekistan aims to reduce GHG emissions per unit of GDP by 35% compared to the level of 2010 by 2030 (Government of Uzbekistan 2021). The NDC contains a set of well-defined and measurable targets:

– double the energy efficiency indicator and reduce the carbon intensity of the GDP.
– bring the share of renewable energy to 25% of total power generation.
– upgrade the infrastructure of industrial enterprises and ensure their sustainability by increasing energy efficiency by at least 20%.

- expand the production and use of motor fuels and vehicles with improved energy efficiency and environmental performance, as well as develop electric transport.
- significantly increase water use efficiency in all sectors of the economy, introduce drip irrigation technologies on up to 1 million hectares of land and increase crop yields cultivated on this land by up to 20–40%.
- achieve Land Degradation Neutrality.
- increase the average productivity of basic agricultural products by 20–25%.

Uzbekistan also mentions in its NDC that it is considering the introduction of a carbon tax mechanism. Compared to its neighbours, Uzbekistan places more emphasis on economic development in its plans for improving energy efficiency and investing in renewable energy production.

3.5 Summary of NDCs

There is considerable diversity in the GHG targets of Central Asian countries. For example, the countries vary in their choice of base year (see Table 2). Kazakhstan and Tajikistan still use 1990 as a baseline, while Kyrgyzstan uses a business-as-usual scenario and Uzbekistan uses 2010 as a base year. The use of 1990 makes reaching GHG emission targets relatively easy since, during the Soviet period, Central Asia was part of a large industrial system with high emission industries.

The foci of the NDCs are also different. For example, Kyrgyzstan is focused on GHG emission reduction and provides detailed estimates of the decrease of emissions for each sector. Strikingly, Kyrgyzstan mostly intends to reduce its emissions through the expansion of the gas supply network. Tajikistan is more focused on adaptation and increasing resilience. Uzbekistan is focused on increasing the carbon efficiency of its economy and ensuring that climate goals do not create impediments to national development programmes. Interestingly, Uzbekistan's NDC explicitly states that the country does not aim to reduce absolute GHG emissions in terms of quantity but instead aims to reduce the carbon intensity of its GDP.

4 National Development Programmes

4.1 Kazakhstan

The 'Strategic Development Plan of the Republic of Kazakhstan until 2025' is the main development programme for the country. Kazakhstan recognizes its dependence on fossil fuels and underlines the importance of the development of non-oil sectors (Government of Kazakhstan 2018). The main goal of the strategy is to increase the productivity and diversification of the economy through technological upgrades and digitalization.

Table 2 Key features of the Central Asian NDCs

Country and year of most recent NDC submission	Baseline	GHG emission reduction target by 2030	Renewable energy target	Level of ambition	Likelihood of achieving goals based on historical trends and current challenges
Kazakhstan, 2016	1990	15%[a] 25%[b]	Not stated	Ambitious	Moderate
Kyrgyzstan, 2021	Business as usual scenario	15.97%[a] 43.62%[b]	Not stated	Highly ambitious	Moderate
Tajikistan, 2021	1990	Not exceed 60–70%[a] of 1990s emission level Not exceed 50–60%[b] of 1990s emission level	Not stated	Unambitious	High
Uzbekistan, 2021	2010	35% per unit of GDP by 2030	25% (including small hydro)	Unambitious	High

Note [a]Unconditional and [b]Conditional upon external support

The strategy has two sections relevant to the NDC: the initiative on 'Promoting technology transfer' and initiative on 'Development of green technologies.' The first initiative aims to upgrade the industrial sector with modern global technology to increase the efficiency of the economy. The second initiative explicitly aims to introduce energy-efficient and smart technologies that are in alignment with climate policy goals. This part of the strategy emphasizes the development of renewable energy and decommissioning of old power plants. The strategy also stipulates that national and international financial resources will be mobilized to support green energy investments, including via technology transfer projects and support for the private sector in energy transition.

4.2 Kyrgyzstan

The general goal of the 'National Development Programme of the Kyrgyz Republic until 2026' is to improve the well-being of citizens by creating a favourable environment for socio-economic development (Government of Kyrgyzstan 2021a).

The programme has a special section titled 'Environmental sustainability and climate change.' This part of the programme stipulates the implementation of 'green economy' principles in sectoral policies and mentions two major areas: increase in energy efficiency and the use of renewable energy. The rest of the section is devoted to conservation of ecosystems and biodiversity, as well as to minimizing environmental damage.

The section also contains eleven dedicated projects which include the development of 'green economy' standards, low-emission public transport, biodiversity conservation, forestry, emergency resilience and re-cultivation of mining sites. However, most of the projects are at the subnational level and are unlikely to have a major impact on climate policy in the country. Despite the promising title, the section fails to incorporate the announced climate goals and does not demonstrate ambition in climate change adaptation and mitigation.

4.3 Tajikistan

'The National Development Strategy of The Republic of Tajikistan until 2030' has four primary objectives: (a) ensuring energy security and efficient use of electricity; (b) turning the country into a major transit country; (c) ensuring food security and access of the population to quality nutrition; (d) expansion of productive employment (Government of Tajikistan 2016).

To ensure energy security, Tajikistan aims to diversify its energy supply, including through hydro energy expansion, modernization of existing power plants, installation of new oil and coal powered plants, use of solar and wind energy and improvements to energy efficiency and saving. The strategy states that the share of electricity generated from coal, oil, gas and renewable sources must be increased by 10%. Tajikistan, in its development programme, clearly intends to develop a fossil-fuel-based energy sector with a strong focus on the modernization of its energy infrastructure.

4.4 Uzbekistan

'The Development Strategy for the New Uzbekistan for 2022–2026' has seven priorities. These include: (1) human dignity and free civil society; (2) rule of law; (3) accelerated economic growth; (4) fair social policy and human capital development; (5) spiritual development; (6) local solutions for global challenges; and (7) national security. In total the strategy has 94 specific and measurable goals (Government of Uzbekistan 2022).

The goal relevant to the NDC is Objective No. 24, 'Uninterrupted supply of electricity to the economy, active introduction of "green economy" technologies in all areas, increase the energy efficiency of the economy by 20 percent.' This involves a whole range of measures related to climate policy, such as raising the share of renewable energy in total energy production to 25% by 2030; improving resource efficiency in the industrial sector; widespread adoption of renewable energy sources;

improvement of energy efficiency in housing; and production and use of electric vehicles.

We can conclude that Uzbekistan's national development strategy is in alignment with its NDC. Both documents focus on energy efficiency improvement in the short run and building a basis for renewable energy generation and expansion.

5 Discussion and Conclusion

Climate change creates many risks for the economies of the Central Asian region, and successful adaptation efforts will require effective policies and investment in many sectors. Central Asian countries have submitted their intentions under the Paris Agreement via their NDCs.

Kazakhstan, Kyrgyzstan, Tajikistan and Uzbekistan are not major contributors to global GHG emissions, but have large climate change mitigation potential through the expansion of renewable energy. They have recently made a number of pledges and declared notable climate policy ambitions. This chapter looked at how well their climate goals are aligned with their national development programmes (Table 3).

The Central Asian energy systems were built during Soviet times and the governments are currently striving to transform the national energy systems to meet the growing energy demand. Historically, fossil fuels have been the main source of energy supply, and in recent years their share has even increased in Kazakhstan, Kyrgyzstan and Tajikistan. Kyrgyzstan has been witnessing a fundamental change in its energy system, where residential consumption became the largest share and largest emitter of GHGs starting from the early 2000s. Facing climate change and crumbling old national energy systems, the governments of the region have announced new policies to upgrade energy supply and improve energy efficiency.

The declared national development plans are siloed by sector and often ignore the findings of climate change impact research. The best example is plans to invest significantly into hydro energy despite the mounting research on looming water stress in the region.

Despite the increasing climate changed-induced water-related risks, the two upstream countries—Kyrgyzstan and Tajikistan—in their national development

Table 3 Summary table

Country	Historical trends (share of RE, emissions, energy intensity) in relation to climate goals	NDC ambition	National development plan alignment with NDC
Kazakhstan	Positive	Ambitious	Generally aligned
Kyrgyzstan	Negative	Highly ambitious	Weakly aligned
Tajikistan	Negative	Low ambition	Not aligned
Uzbekistan	Mixed	Low ambition	Aligned

plans are still heavily focused on the development of new hydropower plants, including both large national dams and small hydro stations. The national energy strategies repeatedly mention the underutilized hydro energy potential and neglect the latest climate change science warnings on water disruptions.

In addition, Kyrgyzstan and Tajikistan seem to be heavily reliant on external support, while Kazakhstan and Uzbekistan have started to deploy renewable energy, set carbon prices and integrate their climate targets into sectoral programmes.

The climate goals of the countries of Central Asia are difficult to compare because they use different base years, metrics and policy measures. Kazakhstan and Tajikistan use 1990 as their base year, Kyrgyzstan uses a business-as-usual scenario and Uzbekistan uses 2010 as its base year.

Kazakhstan has not updated its NDC, yet the country can be considered the region's climate policy leader with actionable climate policy such as intended carbon neutrality by 2060, investment in renewable energy, and the introduction of an emissions trading scheme. With the exception of Kazakhstan, none of the countries has a carbon pricing mechanism, although Uzbekistan's NDC does mention the possibility of introducing carbon pricing.

Kazakhstan and Uzbekistan have taken the first steps towards embracing renewable energy and new technologies while the region's two major hydropower producers—Kyrgyzstan and Tajikistan—still hope to realize their vast hydropower potential. However, the historical trends, NDCs and national development plans for these latter two countries call for an increasing reliance on fossil fuels.

We can thus conclude that Central Asian countries have embraced the global climate change agenda through their signing of the Paris Agreement and declaration of national climate policy targets; however, the countries each rely on different approaches in reaching these targets. Kyrgyzstan and Tajikistan see the GHG reduction obligations as supplementary to their national development, while Kazakhstan and Uzbekistan view their climate commitments as complimentary to the growth of the national economies.

All four countries rely heavily on international partners for financing and technology transfer and have stated that they require such transfer to achieve their climate goals. The current trends and national policies do not suggest that transformational changes in terms of energy transition and GHG emission reduction will occur in the foreseeable future.

References

Bekchanov M, Ringler C, Bhaduri A, Jeuland M (2015) How would the Rogun Dam affect water and energy scarcity in Central Asia? Water Int 40(5–6):856–876. https://doi.org/10.1080/02508060.2015.1051788

Government of Kazakhstan (2016) Intended Nationally Determined Contribution—submission of the Republic of Kazakhstan

Government of Kazakhstan (2018) Strategic Development Plan of the Republic of Kazakhstan until 2025 (No. 636). Art. 636

Government of Kyrgyzstan (2021a) National Development Program of the Kyrgyz Republic until 2026

Government of Kyrgyzstan (2021b) Updated Nationally Determined Contribution of the Kyrgyz Republic

Government of Tajikistan (2016). National Development Strategy of the Republic of Tajikistan until 2030

Government of Tajikistan (2021) The updated NDC of the Republic of Tajikistan

Government of Uzbekistan (2021) Updated Nationally Determined Contribution of the Republic of Uzbekistan

Government of Uzbekistan (2022) Development strategy for the new Uzbekistan for 2022–2026

IEA (2022) World energy balances, April 17

IPCC (2022) Climate change 2022: impacts, adaptation and vulnerability

Koch N, Tynkkynen V-P (2021) The geopolitics of renewables in Kazakhstan and Russia. Geopolitics 26(2):521–540. https://doi.org/10.1080/14650045.2019.1583214

Laldjebaev M, Morreale SJ, Sovacool BK, Kassam K-AS (2018) Rethinking energy security and services in practice: national vulnerability and three energy pathways in Tajikistan. Energy Policy 114:39–50. https://doi.org/10.1016/j.enpol.2017.11.058

Poberezhskaya M, Bychkova A (2021) Kazakhstan's climate change policy: reflecting national strength, green economy aspirations and international agenda. Post-Communist Econ: 1–22. https://doi.org/10.1080/14631377.2021.1943916

Reyer CPO, Otto IM, Adams S, Albrecht T, Baarsch F, Cartsburg M, Coumou D, Eden A, Ludi E, Marcus R, Mengel M, Mosello B, Robinson A, Schleussner C-F, Serdeczny O, Stagl J (2017) Climate change impacts in Central Asia and their implications for development. Reg Environ Change 17(6):1639–1650. https://doi.org/10.1007/s10113-015-0893-z

Sabyrbekov R, Ukueva N (2019) Transitions from dirty to clean energy in low-income countries: insights from Kyrgyzstan. Cent Asian Surv 38(2):255–274. https://doi.org/10.1080/02634937.2019.1605976

Vakulchuk R (2022) Energy security and green energy in Central Asia. In: Central Asian Bureau for Analytical Reporting (CABAR).

Vakulchuk R, Isataeva A, Kolodzinskaia G, Overland I, Sabyrbekov R (2022) Fossil fuels in Central Asia: trends and energy transition risks. Cent Asia Reg Data Rev 28:1–6. https://doi.org/10.13140/RG.2.2.11461.37607

World Bank (2017) Analysis of the Kyrgyz Republic's energy sector. In: World Bank energy and extractives global practice ECA region, May. https://doi.org/10.1596/29045

World Bank (2022) World Development Indicators

Rahat Sabyrbekov is an environmental economist who specializes in decarbonization, climate change and energy transition. Rahat is a Postdoctoral Fellow at OSCE Academy and Visiting Fellow at Davis Center at Harvard University. He obtained his PhD from School of Economics and Business at Norwegian University of Life Sciences. He received his Master's degree from University of Birmingham, the United Kingdom. He teaches Economics of Sustainable Management of Mineral Resources course at the OSCE Academy. His recent publications include *Putting the Foot Down: Accelerating EV Uptake in Kyrgyzstan* (2023), *Know Your Opponent: Which Countries might Fight the European Carbon Border Adjustment Mechanism?* (2022), *Fossil Fuels in Central Asia: Trends and Energy Transition Risks* (2022).

Indra Overland is Research Professor and Head of the Research Group on Climate and Energy at the Norwegian Institute of International Affairs (NUPI). He previously headed the Russia and Eurasia Research Group at NUPI and has worked on Central Asia since 2001. He completed his PhD at the University of Cambridge, followed by a three-year post-doctoral project on Central Asia and the South Caucasus. He has carried out fieldwork in all Central Asian states and has been responsible for cooperation between the OSCE Academy in Kyrgyzstan and NUPI since 2007.

Every year he teaches MA students from all Central Asian countries on energy issues and hosts 4-5 Central Asian students in Norway. He is (co)author of *Caspian Energy Politics* (Routledge), *China's Belt and Road Initiative Through the Lens of Central Asia* (Routledge), *Kazakhstan: Civil Society and Natural Resource Policy in Kazakhstan* (Palgrave) and *Renewable Energy Policies of the Central Asian Countries* (CADGAT). He is a contributing author to the IPCC's Sixth Assessment Report (Working Group II).

Roman Vakulchuk is Senior Researcher at the Norwegian Institute of International Affairs (NUPI) in Oslo and holds a PhD degree in Economics from Jacobs University Bremen in Germany. He specializes in Central Asia and Southeast Asia and his main research interests are economic transition, trade, energy, climate change and investment policy. Vakulchuk has served as project leader in research projects organized by the Asian Development Bank (ADB), the World Bank, the Global Development Network (GDN), the Natural Resource Governance Institute (NRGI) and others. In 2018, he worked as governance expert for OECD's mission in Kazakhstan and advised the government on privatization reform. In 2013, Vakulchuk was awarded the Gabriel Al-Salem International Award for Excellence in Consulting. Recent publications include *Seizing the Momentum. EU Green Energy Diplomacy towards Kazakhstan* (2021), *Discovering Opportunities in the Pandemic? Four Economic Response Scenarios for Central Asia* (2020) and *Renewable Energy and Geopolitics: A Review* (2020).

Open Access This chapter is licensed under the terms of the Creative Commons Attribution 4.0 International License (http://creativecommons.org/licenses/by/4.0/), which permits use, sharing, adaptation, distribution and reproduction in any medium or format, as long as you give appropriate credit to the original author(s) and the source, provide a link to the Creative Commons license and indicate if changes were made.

The images or other third party material in this chapter are included in the chapter's Creative Commons license, unless indicated otherwise in a credit line to the material. If material is not included in the chapter's Creative Commons license and your intended use is not permitted by statutory regulation or exceeds the permitted use, you will need to obtain permission directly from the copyright holder.

Decarbonisation Opportunities and Emerging Carbon Pricing Instruments in Central Asia

Gulim Abdi, Nurkhat Zhakiyev, and Shynar Toilybayeva

Abstract Central Asian countries are highly vulnerable to climate change and heavily reliant on fossil fuel resources. All countries need to decarbonise their economies for sustainable growth and to meet their Paris Agreement goals. Within this global challenge, there are significant opportunities for Central Asian countries, such as attracting green investment through the expansion of renewable energy, phasing out fossil fuel subsidies and improving energy efficiency. This chapter presents the results of a study that investigated how carbon pricing instruments are currently used in Central Asia and what the future holds. The study determined that CPI and decarbonisation strategies are now being considered by Central Asian countries at different levels. A SWOT analysis of carbon pricing instrument implementation revealed ways to facilitate the implementation of carbon pricing to decarbonise the regional economy. The study also identified a list of mid- and long-term decarbonisation activities for governments and other stakeholders, and analysed opportunities to strengthen regional cooperation.

Keywords Decarbonisation · Central Asia · Carbon pricing · Emissions trading · Climate change

1 Introduction

The purpose of this chapter is to provide a regional overview of the carbon pricing trends in Central Asia and to propose strategic recommendations for decarbonisation. Many different trade-offs and adjustments to the energy mix can be made if new decarbonisation initiatives are introduced to the region's energy systems.

G. Abdi · N. Zhakiyev (✉)
Astana IT University, Astana, Kazakhstan
e-mail: nurkhat.zhakiyev@astanait.edu.kz

S. Toilybayeva
CAREC Country Office, Astana, Kazakhstan

© The Author(s) 2023
R. Sabyrbekov et al. (eds.), *Climate Change in Central Asia*,
SpringerBriefs in Climate Studies,
https://doi.org/10.1007/978-3-031-29831-8_5

Setting a price on greenhouse gas (GHG) emissions, either directly through a carbon tax or indirectly through an emissions trading system (ETS), is one of the most important policy interventions for reducing GHG emissions. Carbon pricing instruments (CPIs) can stimulate a transition to a low-carbon economy (UNFCCC 2021a), and CPIs are being rolled out around the world as effective policy tools. Since 2012, the overall number of CPIs has nearly doubled (World Bank 2020). Yet despite the progress that has been made to date, approximately 78% of global carbon emissions are not yet priced. Existing carbon pricing schemes have a much lower tariff than the 40–80 dollars per tonne thought to be necessary to meet nationally determined contribution (NDC) goals and the new challenges discussed at the 2021 United Nations Climate Change Conference in Glasgow (COP 26) (World Bank 2021; Nguyen 2019).

Beyond the observation that carbon pricing is being adopted as a policy tool, little is known about key barriers to or issues in implementing market-oriented mechanisms in the Central Asia region (Vakulchuk et al. 2022a). To fill this gap, we carried out a comparative review analysis to investigate adoption tendencies (relating to both current and planned policies) with a view to ascertaining what significant opportunities can be identified for the decarbonisation of the region's economy.

2 Regional Overview

2.1 *Emissions Profile and Fossil Fuel Subsidies*

Kazakhstan, Uzbekistan and Turkmenistan have the largest carbon footprints among the Central Asian countries (UNFCCC 2022). In total, the region emitted 710.5 million tonnes of CO_2-equivalent ($MtCO_2e$) in 2019, with Kazakhstan accounting for 55.7% (~396 $MtCO_2e$), Uzbekistan 28.9% (~205 $MtCO_2e$), Turkmenistan 12% (~85 $MtCO_2e$), Kyrgyzstan 2.1% (~15 $MtCO_2e$) and Tajikistan 1.3% (~9.5 $MtCO_2e$). The difference between the countries is thus significant (World Bank 2022). It is also worth noting that emissions for the region peaked in the 1990s, before the dissolution of the Soviet Union, and have not reached such levels since.

Kazakhstan and Turkmenistan are substantial oil and gas exporters whose per capita emissions grew by almost 70% from 1998 to 2014. This is not reflected in Fig. 1 as it relates emissions to purchasing power parity (PPP) rather than the size of the population. During the same period, Uzbekistan's per capita emissions fell by 1.6 tonnes. The economies of the lower-emitting countries, Kyrgyzstan and Tajikistan, rely on agriculture which is exposed to climate change. Those countries—Kazakhstan, Uzbekistan and Turkmenistan—with higher economic growth (based on fossil fuel exports) also have higher energy demands, with high levels of fossil fuel consumption in the power sector. This clearly highlights an urgent need to implement new beneficial CPIs for these economies. Tajikistan, on the other hand, utilises hydropower for 98% of the country's electricity generation (IEA 2020). Yet this is

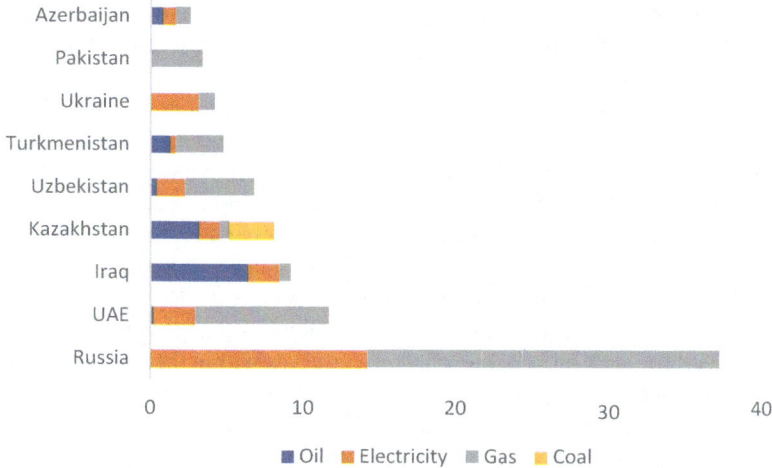

Fig. 1 Fuel consumption subsidies by selected countries (compiled by authors based on International Energy Agency data) (IEA 2018)

only 5% of its hydropower potential. The other Central Asian countries are also using only a fraction of their current hydropower potential (see below). Hydropower would therefore appear to be key to the decarbonisation of the region's power generation (Reyer et al. 2015).

There is also another way to diversify the economies of these countries. The region's heavy reliance on fossil fuel exports could be replaced with a focus on the export of mineral resources for the green transition. If Central Asia's production of mineral resources for green technologies expands, the adjustment would help the region's decarbonisation and strengthen its global position as a critical materials supplier (Vakulchuk and Overland 2021). This is especially true for the major fossil fuel producers.

Kazakhstan, Uzbekistan and Turkmenistan heavily subsidise their domestic consumption of fossil fuels, lowering retail prices of power and transport fuel below the world average and discouraging the expansion of renewable energy (RE) for power generation. Large-scale fossil fuel subsidies are known to undermine the effect of CPIs. Central Asian governments should therefore review their fossil fuel subsidies as carbon pricing methods evolve (IEA 2018) (Fig. 1).

2.2 Climate Commitments and RE Potential

Central Asia has considerable potential for generating electricity from wind, solar and hydropower. Uzbekistan already exploits 40% of its technically feasible hydropower potential, while Kazakhstan and Kyrgyzstan utilise only 13% and 15% respectively and Turkmenistan has only one hydropower station. While hydropower represents 98% of Tajikistan's domestic power supply, the country is still only using 5% of its hydropower potential.

RE must be secured against intermittency. Greater efficiency and grid stability and greater RE penetration might be achieved by integrating diverse energy sources (e.g. gas, water, wind, solar) into unified power generation systems across countries. The individual Central Asian countries have different strengths and seasonal dynamics in power generation. For instance, Tajikistan and Kyrgyzstan have strong hydropower potential in the summer, while Uzbekistan and southern Kazakhstan have good thermal electricity potential in the winter. Yet electricity provision in Central Asia remains divided. Cross-border electricity exchanges in the region could reduce the need for peak and backup capacity reserves within the respective national systems, while optimising electricity transfer between states will optimise the use of water resources, especially in summer. By fully exploiting and further expanding the area's integrated power transmission infrastructure, countries could significantly lower their operational costs. Discussions are being conducted at regional and international levels to increase interconnectivity and improve system value. For instance, the CASA-1000 project, connecting nations in Central Asia (Kyrgyzstan and Tajikistan) and South Asia (Afghanistan and Pakistan), will expand power trading prospects within and beyond the Central Asian region (World Bank 2020). The Central Asian countries may thus increasingly be able to identify opportunities to export green electricity.

2.3 Consideration of CPI Adoption

Kazakhstan is the first (and to date the only) Central Asian country to have introduced a CPI. This is in the form of a domestic emissions trading system (ETS) which was launched as a pilot in 2013 (CAREC 2021). All other Central Asian countries are currently considering carbon pricing tools at various institutional levels to cut emissions in accordance with their NDCs, but the governments have not yet made any formal commitments to CPIs (Table 1).

3 Methods

This chapter employs a comparative review analysis of publicly available information from national reports produced by policymakers in each Central Asian country. The authors also conduct a SWOT analysis of the potential for deploying carbon pricing instruments (CPI) in each country. Individual country profiles are presented first, followed by a discussion of the results of the SWOT analysis and a consideration

Table 1 Adoption and consideration of CPIs by Central Asian countries

Country	CPI status	Other relevant climate policies
Kazakhstan	ETS adopted	• Green Economy concept 2013 (NCTGE 2013) • Carbon tax is considered as a deep decarbonisation scenario in the updated NDC Roadmap for 2021–2030 • Low-Emission Development Strategy (to 2050)
Turkmenistan	ETS or green certificates considered	• National Strategy on Climate Change (2019) • National Green Economy Concept (2020) which considers green certificates
Uzbekistan	ETS considered	• Green Economy Concept (2019) • A carbon price mechanism is one of the priorities in the Carbon Neutral Electricity Sector Programme (2020)
Kyrgyzstan	ETS considered	• Climate Change Adaptation Programme not approved yet • Sustainable Development Strategy of the Kyrgyz Republic for the period of 2018–2040 • NDC roadmap for 2021–2030
Tajikistan	ETS or green certificates considered	• NDC roadmap for 2021–2030 • Climate Change Mitigation and Adaptation Programme (2019) • Programme for supporting hydropower for 2016–2025

of the potential for collaboration between the countries towards regional decarbonisation. On the basis of the analysis, we develop recommendations for governments and other key stakeholders.

4 Carbon Pricing Considerations in Individual Central Asian Countries

In this section we present the findings of the comparative review, highlighting the key factors that influence the success of decarbonisation efforts, including the adoption and effective implementation of CPIs, in individual Central Asian countries. Following the presentation of the individual country profiles, we consider the findings in a cross-regional perspective.

4.1 Kazakhstan

Kazakhstan is Central Asia's largest GHG emitter. In 2018, it produced 396 $MtCO_2e$ of GHG, close to the 1990 baseline. The 'Concept of Transition towards a Green Economy—2050' was adopted in 2013. Renewable sources contributed 3% of electricity generation in 2020 (excluding large hydropower, which accounted for 10%). Kazakhstan proposes to increase the percentage of gas-fired power plants in the overall electricity generation mix to 20% by 2020, 25% by 2030 and 30% by 2050 as a way to gradually reduce the contribution of coal-fired power stations (NCTGE 2013). Gas power with a high ramping rate helps to increase the share of intermittent RE. By 2030, GHG emissions from electricity generation are to be decreased by 15%, and by 40% by 2050 (Yeserkepova et al. 2014). To be in line with the Paris Agreement's objectives, these targets must be significantly increased. It is critical to increase the share of RE in power generation while decreasing the shares of both natural gas and coal. The government understands its significant carbon footprint and its reliance on coal and is taking steps to enhance the share of renewables in the energy mix, expand its carbon ETS, and increase gasification.

Kazakhstan has announced that it aims to achieve carbon neutrality by 2060 (Vakulchuk et al. 2022b). In one of the scenarios outlined in the Fourth Biennial Report of the Republic of Kazakhstan to the UNFCCC (2019), methane emissions are included under the ETS from 2026 and GHG trading auctions are to be instituted, while the proceeds from auctions and fines will go into a national decarbonisation fund (UNFCC 2019). A carbon tax on unregulated industries has also been considered to fulfil the NDC goals (Wong and Zhang 2022). National plans for the allocation of GHG quotas have been adopted in three phases. In Phase 1, the First National Plan was piloted. Phase 2 (2014–2015) began with the allocation of quotas based on historical emissions. In 2015, independent verification companies started to verify the GHG reports. A process of collection, processing and analysis of GHG emissions data was initiated to assess the benchmarking system for allocating quotas. Kazakhstan has now established facility-level measurement, reporting and verification (MRV) systems as the basis for its domestic ETS (Gulbrandsen et al. 2017).

From 2016 to 2018, Kazakhstan suspended the trading of GHG quotas under the ETS owing to distortions in the system and defects that needed to be eliminated (within the historical method). The government also addressed legal conflicts and gaps in carbon regulation that had been discovered during the previous phases (NIR-KZ UNFCCC 2020; UNFCCC 2019). In 2018, the ETS resumed operations after a two-year suspension, with new distribution methods and trading procedures based on a baseline or benchmarking level. Operators are not obliged to report on emissions of GHGs other than CO_2. Emissions of other GHGs account for approximately 17% of the total GHG emissions in Kazakhstan (NIR-KZ UNFCCC 2020).

Phase 3 of the ETS (2018–2020) uses benchmarking and historical quota distribution as the basis for reporting. The National GHG Allocation Plan adopted in January 2018 sets emissions limits for 129 enterprises for the 2018–2020 period, and a 5% emissions reduction target for 2020 (Gulbrandsen et al. 2017).

4.2 Turkmenistan

In their first NDC under the Paris Agreement, submitted in 2016, Turkmenistan explains that the main sources of GHGs in the country are oil and gas enterprises, energy, agriculture and transport industries, as well as housing and communal services. Turkmenistan has the highest energy and carbon intensity per PPP to GDP in the region. Power generation is dominated by natural gas and oil, and significant emissions reductions could be achieved by reducing energy losses in electricity and gas networks and by tackling venting and flaring. The solar and wind potential in the country is very high but underutilised. The development of Turkmenistan during the years following independence was marked by high growth in industrial production and investment in the economy. The country's rapid economic growth is linked to an increase in the use (and export) of energy products, primarily oil and gas, which has contributed to an increase in GHGs. According to emissions estimates over the 18-year period from 2000 to 2017, Turkmenistan's GHG emissions more than doubled (Holzhacker and Skakova 2018).

To reduce emissions, Turkmenistan has developed a National Strategy on Climate Change (2019), identifying three main areas for greener economic development, including energy efficiency and conservation, sustainable use of natural gas and petroleum products and increased use of alternative energy sources. In the National Strategy, one of the key proposed actions is studying international experience and preparing a proposal for the development of a green certificate trading system. This complex task provides recognition of the results of a national inventory conducted by local experts and reputable international organisations (MFA 2019).

In 2017, the President of Turkmenistan signed the Decree 'On Approval of the State Energy Saving Programme for the Period 2018–2024', according to which a National Inventory System for GHGs should be prepared, along with MRV guidelines to ensure transparency in its implementation.

4.3 Uzbekistan

The World Bank categorises Uzbekistan as one of the world's most vulnerable countries to climate change (ClimatePolicyDatabase 2021). Uzbekistan emits around 205 $MtCO_2e$ annually. About 89.4% of the GHG emissions in Uzbekistan are generated by the energy sector in the western part of the country, where the capital Tashkent is located (UNFCCC 2021b).

A number of policy measures have been adopted that will facilitate Uzbekistan's transition to low-carbon energy including the National Green Economy Strategy for the period to 2030. This strategy requires an increase in the share of RE in electricity production to at least 25% by 2030. Currently, Uzbekistan is developing legislation in the field of RE, taking into consideration the experience of developed countries and the country's growing need for energy. The country has also revised building codes and regulations to bring them in line with higher energy efficiency standards. As a result, almost 10 GW will be added to the country's RE capacity, including 3 GW of wind turbines, 1.9 GW from hydroelectric power plants and 5 GW of solar power (ClimatePolicyDatabase 2021; Uzbekistan's NDC 2017). The latter excludes the capacity of individual households.

While these developments are encouraging for Uzbekistan's RE growth, legislative deficiencies prevent the government from adopting a sufficiently ambitious, comprehensive and long-term strategy to decarbonise its energy sector. One of the main goals adopted by Uzbekistan was to reduce specific GHGs per unit of GDP by 10% by 2030. To do so the government plans to expand the introduction of environmental protection measures, strengthen legislation in the field of ecology and introduce the appropriate world standards for equipping newly built enterprises. Beyond the existing 10% reduction target from 2010 levels, more ambitious targets are possible (Uzbekistan's NDC 2017).

In 2021, Uzbekistan published its Roadmap for a Carbon Neutral Electricity Sector by 2050, which provides a framework for decarbonising the country's electricity sector. The proposed transition in power generation would require considerable technological and regulatory revisions with strong political support. The roadmap states that Uzbekistan must eliminate legal and institutional preferences for carbon-intensive sources and eventually establish a carbon pricing system to assist in phasing out fossil fuel subsidies in favour of renewables and green economy initiatives. However, without a fully liberalised electricity market where renewables can compete and deliver both cost-related and environmental benefits to consumers, carbon pricing is unfeasible for operators (EBRD 2021).

Converting gas into more valuable products can also result in greater economic benefits, jobs and investment (Uzbekistan's NDC 2017; EBRD 2021). For instance, Uzbekistan could also use domestic gas supplies and extra RE output in the development of a hydrogen economy. Further areas to explore include the potential to reduce methane emissions from the oil and gas sector and lower nitrous oxide and fluorinated emissions from industrial operations (EBRD 2021).

Uzbekistan has designed MRV systems as part of its Nationally Appropriate Mitigation Actions initiatives, completing an initial evaluation and developing a major road map. CPI in Uzbekistan exists currently only in concept form.

4.4 Kyrgyzstan

Kyrgyzstan recognises its vulnerability to climate change, but also that its planned economic development will lead to a significant increase in GHG emissions if they are unabated. In February 2020, Kyrgyzstan submitted its first NDC under the Paris Agreement. In it, Kyrgyzstan commits to cutting GHGs by 11.49–13.75% below the 'business-as-usual' scenario (BAU) by 2030 (Takeuchi 2020). With international cooperation, Kyrgyzstan plans to implement mitigation measures to reduce GHGs by

29–30% below BAU by 2030. The country's long-term unconditional contribution is to reduce GHGs by 12.67–15.69% below BAU by 2050.

Kyrgyzstan's seemingly low level of emissions is not actually that low if one takes into account that 90% of the country's electricity is hydropower (National Statistical Committee of the Kyrgyz Republic 2020). However, climate change is likely to have an impact on the water supply, reducing the potential of hydropower at the same time as electricity demand is expected to grow significantly (Temirbekov 2020).

Considering the country's emissions profile and Paris Agreement goals, the government is investigating carbon pricing mechanisms. However, there are several barriers, including a lack of conceptual understanding of the benefits of carbon pricing, as well as political will and supportive legal and fiscal frameworks. There is also a lack of capacity, technical understanding, and knowledge of products for CPI implementation. There is also no MRV system, as Kyrgyzstan's emissions reduction program is in the initial assessment stage.

Owing to these barriers, Kyrgyzstan has yet to consider carbon pricing. There are other potential barriers to the implementation of CPI that include companies perceiving it as restrictive to growth and consumers viewing it as a cost that could lead to poverty. Carbon price increases could thus potentially cause social tension if not handled correctly (Proskuryakova and Ermolenko 2022).

4.5 Tajikistan

In Tajikistan, climate mitigation and adaptation actions are reflected in the National Development Strategy to 2030, adopted in 2019 (UNFCCC 2021a), and in the country's medium-term development program for 2021–2025. Tajikistan submitted its first NDC under the Paris Agreement in 2017, with the emission reduction targets summarised in Table 2.

Compared to the 1990 baseline, GHG emissions in Tajikistan had decreased by 64.3% by 2014 and amounted to 9.1 $MtCO_2e$. This is largely due to the heavy utilisation and expansion of hydropower, which replaced coal-fired power stations and as of 2020 accounted for 98% of the country's electricity generation (IEA 2020). Tajikistan's first Biennial Report was submitted in 2019, the process of preparing the fourth national communication under the UNFCCC is currently underway. Tajikistan's carbon pricing scheme is in its early stages. However, according to official

Table 2 Tajikistan's conditional and unconditional NDC targets to 2030

#	Conditions	Reduction potential from 1990 level (25.5 $MtCO_2e$)	Targeted specific emissions per capita, tCO_2e per capita
1	Unconditional target by 2030	Minus 10–20% (20.4–23 $MtCO_2e$)	1.7–2.2
2	Conditional target by 2030 (with international support)	Minus 25–35% (16.6–19.1 $MtCO_2e$)	1.2–1.7

publications, the country expects to develop an MRV legislative framework by 2025 and a pricing system by 2030 (UNFCCC 2021a). In this regard, the country's organisational, financial, informational and human capabilities are major barriers. Tajikistan will require international financial support to establish its MRV system and may encounter difficulty with data collection.

5 Discussion

According to the results of the comparative review analysis, there are significant opportunities for decarbonisation and CPI within the Central Asia region. The region's energy infrastructure needs to be modernised to decrease grid and heat transmission losses while increasing energy generation and distribution efficiency. The countries, especially Kazakhstan, Uzbekistan and Turkmenistan, all have very significant fossil fuel subsidies, distorting the cost of energy and fuels and resulting in increased GHG emissions. Fossil fuel subsidies encourage overconsumption of fossil fuels and discourage the adoption of RE technologies. Retail electricity rates are also below the global average in all Central Asian countries (IEA 2020). Energy and resource efficiency must be improved if the countries want to establish internationally competitive manufacturing and service sectors.

Central Asia has great potential for the deployment of RE, in particular through the integration of regional electricity grids. Increased collaboration and coordination among the Central Asian countries might significantly boost RE penetration at lower costs, while also optimising grid stability. Central Asia's power system modelling shows significant operating cost savings in the next decade. Discussions are currently ongoing to increase interconnectivity and maximise system value, including as part of the CASA-1000 cross-regional power connectivity project (World Bank 2020; Otarov et al. 2017). In this context, the region may increasingly find opportunities to export green electricity generated by RE. Meanwhile, all Central Asian countries are exchanging commitments under the Paris Agreement and setting up long-term low to net zero emission development strategies, showcasing their long-term planning and forming a potential basis for regional collaboration.

5.1 Decarbonisation Opportunities and CPI: A SWOT Analysis

Despite the growing global trend in the adoption of CPIs to achieve NDCs cost-effectively, Kazakhstan is the first and only Central Asian mover in carbon pricing. Although other Central Asian countries have yet to adopt a CPI, they are highly interested in carbon pricing approaches and in potentially reaping the benefits of regional collaboration.

Table 3 SWOT Analysis of CPI in Central Asia

Strengths	Weaknesses
– Governments are already providing support for decarbonisation via programs and projects – Existing MRV tools considered and can be adopted – There are ongoing cooperation projects through regional platforms – The major emitting sectors in all countries are suitable for ETS	– CPI is not aligned with national policies and initiatives in some countries – Limited experience of CPI – Insufficient government support for CPI adoption (apart from in Kazakhstan) – Insufficient technical information in Russian and other Central Asian languages – Lack of local experts with sufficient technical expertise to develop and implement CPI programs and initiatives
Opportunities	Threats
– Less competition among state and private organisations to implement CPI than in some other countries – Existing support structures from international organisations, including the UN, World Bank, USAID, EBRD, ADB and GIZ – Significant GHG mitigation opportunities – Opportunities to diversify economies and create new sources of export – Opportunity to cut fossil fuel subsidies – Significant potential for increased deployment of RE – Synergies between regional electricity cooperation and regional ETS could kill two birds with one stone	– Emissions trading costs would be an additional cost for businesses/consumers, and might not be welcomed – Unless efficient compensation systems for vulnerable groups are established and funded by carbon pricing revenues, price increases (e.g. in electricity or fuel) may lead to social tension – CPI is perceived as an additional burden on the government – Trade relations with countries outside of Central Asia might be impacted by CPI

Our SWOT analysis (Table 3) seeks to identify existing strengths, weaknesses, opportunities and threats that could—or already do—support or undermine the adoption of CPI in Central Asia. The aim is to highlight areas for individual countries and the region as a whole to focus their efforts on, thus increasing the adoption of CPIs throughout the region. The SWOT analysis was based on the findings from the comparative review analysis.

As shown in Table 3, there are still substantial weaknesses and threats to the adoption of CPIs in Central Asia, but there are also considerable strengths and opportunities that might enable their adoption and implementation.

5.2 Recommendations: Actions to Enhance Regional Cooperation on Decarbonisation

On the basis of the comparative review and SWOT analysis, we have developed a set of recommendations for Central Asian governments and key stakeholders to strengthen individual country performance and regional cooperation on decarbonisation and promote the adoption of carbon pricing. We have divided these actions into short-term and longer-term actions.

Short-term actions

1. Central Asian governments should build political will among themselves for the introduction of innovative climate policy tools, including CPI, develop technological regulations for GHG emission inventory to support the implementation of NDCs, national green economy and sustainable development strategies, and green COVID-19 recovery plans.
2. Governments can adopt more ambitious climate commitments in their NDCs, including a willingness and readiness to use new market mechanisms under Article 6 of the Paris Agreement, which allows for the trade of emission reduction units between countries.
3. Governments should support research institutions in examining the implications of phasing out fossil fuel subsidies made by the government in favour of RE and other green economy initiatives.
4. International financial institutions and governments should focus investments on regional cross-border electricity exchange to reduce peak and backup capacity requirements, enable seasonal hydropower trading, reduce required capacity reserves within national systems and promote the growth of RE.
5. The global community must offer for the consideration and potential use of cooperative approaches at the regional and international levels under Article 6 of the Paris Agreement and applicable provisions.
6. Stakeholders must develop and adopt facility-level MRV guidelines for CPI in applicable sectors.

6 Conclusion

According to the results of the comparative review, Kazakhstan, Uzbekistan and Turkmenistan still produce a large amount of GHG emissions. One of the main reasons is their significant subsidies for fossil fuels, which distort the cost of energy and fuel, encourage excessive consumption, and create barriers to the introduction of renewable energy technologies.

This study concludes that all the Central Asian countries can benefit from an economic strategy based on a decarbonisation policy, achieve sustainable growth decoupled from GHG emissions and meet their NDC commitments. Encouraging RE development, reducing fossil fuel subsidies, and increasing energy efficiency and foreign green investment will be the key drivers for decarbonisation in the region.

Central Asian countries can utilise CPIs to stimulate sustainable economic growth and green investment and enable deep decarbonisation in the longer term. Indirect regulatory and institutional support for carbon-intensive sources should be gradually reduced to allow a fair market in the region. Energy subsidies need to be phased out and replaced with cost-reflective tariffs. Regional ETS and phased-out fossil fuel subsidies may thus be considered in the long run to stimulate low-carbon development in the region.

Acknowledgements Co-authors N.Z. and G.A. acknowledge the Ministry of Science and Higher of the Republic of Kazakhstan (Grant No. AP09261258). N.Z. also acknowledges the Scholarship Programme of the Islamic Development Bank (ID 2021-588606).

References

CAREC (2021) The UNFCCC activity in the Central Asian region. https://carececo.org/en/main/news/. Accessed 14 Dec 2021
ClimatePolicyDatabase.org (2021) Uzbekistan-policy database. http://climatepolicydatabase.org/index.php/Country:Uzbekistan. Accessed 14 Jan 2022
EBRD (2021) A carbon neutral electricity sector in Uzbekistan. http://minenergy.uz/en/lists/view/131. Accessed 14 Dec 2021
Gulbrandsen LH, Sammut F, Wettestad J (2017) Emissions trading and policy diffusion: complex EU ETS emulation in Kazakhstan. Glob Environ Polit 17(3):115–133. https://doi.org/10.1162/glep_a_00418
Holzhacker H, Skakova D (2018) Turkmenistan diagnostic. Report. EBRD, London. https://www.ebrd.com/documents/policy/country-diagnostic-paper-turkmenistan.pdf. Accessed 14 Dec 2021
IEA (2018) Fossil-fuel consumption subsidies by country, Paris. https://www.iea.org/data-and-statistics/charts/fossil-fuel-consumption-subsidies-by-country-2018. Accessed 14 Dec 2021
IEA (2020) Shaping a secure and sustainable energy future for all. https://www.iea.org/. Accessed 14 Dec 2021
MFA (2019) Adoption of the national strategy of Turkmenistan on climate change. https://www.mfa.gov.tm/index.php/en/news/1615. Accessed 13 Jan 2021
National Statistical Committee of the Kyrgyz Republic (2020) Monitoring of the Sustainable Development Goal indicators in the Kyrgyz Republic. http://www.stat.kg/media/files/0df67c73-a648-4c10-a6b5-c4df2311cac7.pdf. Accessed 14 Dec 2021
Nguyen AT (2019) The relationship between economic growth, energy consumption and carbon dioxide emissions: evidence from central Asia. Eurasian J Bus Econ 12(24):1–15
NIR-KZ UNFCCC (2020) National inventory report of Kazakhstan on the anthropogenic greenhouse gas emissions for 1990–2018. https://unfccc.int/documents/253715. Accessed 14 Dec 2021
NCTGE (2013) National concept for transition to a green economy up to 2050. Approved by Decree of the President of the Republic of Kazakhstan on May 30 2013. N 557. https://strategy2050.kz/en
Otarov, R, Akhmetbekov Y, Zhakiyev N (2017) Determination of optimal CO_2 allowance prices for stimulation of investments in CCS, RES and other carbon-clean technologies in Kazakhstan. In: Sustainable energy in Kazakhstan. Routledge, pp 123–133. https://doi.org/10.4324/9781315267302-9
Proskuryakova L, Ermolenko G (2022) Decarbonization prospects in the commonwealth of independent states. Energies 15(6):1987
Reyer CPO, Otto IM, Adams S et al (2015) Climate change impacts in Central Asia and their implications for development. Reg Environ Chang 17(6):1639–50. https://doi.org/10.1007/s10113-015-0893-z

Takeuchi K (2020) Project information document-integrated safeguards data sheet-enhancing resilience in Kyrgyzstan additional financing-P172761

Temirbekov A (2020) Nationally Determined Contributions of the Kyrgyz Republic to the Paris Agreement. http://aarhus.kg/wp-content/uploads/2020/11/3.-Temirbekov-ONUV.pdf

UNFCCC (2017) Third National Communication of the Kyrgyzstan. https://unfccc.int/sites/default/files/resource/NC3_Kyrgyzstan_English_24Jan2017_0.pdf. Accessed 14 Dec 2021

UNFCCC (2019) Fourth biennial report of the republic of Kazakhstan to the un framework convention on climate change. https://unfccc.int/sites/default/files/resource/BR4_en.pdf. Accessed 14 Dec 2021

UNFCCC (2021a) The updated NDC of Tajikistan. https://www4.unfccc.int/sites/ndcstaging/PublishedDocuments/Tajikistan%20First/NDC_TAJIKISTAN_ENG.pdf. Accessed 4 Jan 2021

UNFCCC (2021b) Uzbekistan Biennial update report (BUR). https://unfccc.int/sites/default/files/resource/FBURUZeng.pdf. Accessed 5 Jan 2021

UNFCCC (2022) About carbon pricing, May. https://unfccc.int/about-us/regional-collaboration-centres/the-ci-aca-initiative/about-carbon-pricing. Accessed 14 Dec 2021

Uzbekistan's NDC (2017) Uzbekistan's Nationally Determined Contributions. https://www4.unfccc.int/sites/ndcstaging/PublishedDocuments/Uzbekistan%20First/INDC%20Uzbekistan%2018-04-2017_Eng.pdf. Accessed 14 Jan 2021

Vakulchuk R, Overland I (2021) Central Asia is a missing link in analyses of critical materials for the global clean energy transition. One Earth 4(12):1678–1692. https://doi.org/10.1016/j.oneear.2021.11.012

Vakulchuk R, Daloz AS, Overland I, Sagbakken HF, Standal K (2022a) A void in Central Asia research: climate change. Cent Asian Surv 42:1–20. https://doi.org/10.1080/02634937.2022.2059447

Vakulchuk R, Isataeva A, Kolodzinskaia G, Overland I, Sabyrbekov R (2022b) Fossil fuels in Central Asia: trends and energy transition risks. Cent Asia Regi Data Rev 28. https://www.researchgate.net/publication/357866530

Wong JB, Zhang Q (2022) Impact of carbon tax on electricity prices and behaviour. Financ Res Lett 44:102098. https://doi.org/10.1016/j.frl.2021.102098

World Bank (2020) Central Asia electricity trade brings economic growth and fosters regional cooperation 2020. https://www.worldbank.org/en/news/feature/2020/10/20/central-asia-electricity-trade-brings-economic-growth-and-fosters-regional-cooperation. Accessed 14 Dec 2021

World Bank (2021) State and trends of carbon pricing 2020. World Bank, Washington, DC, May. https://doi.org/10.1596/978-1-4648-1586-7

World Bank (2022) National CO_2 emissions of CA countries 1992–2016. The World Bank Group. https://data.worldbank.org/indicator/EN.ATM.CO2E.PP.GD?end=2016&locations=KZ-UZ-TM-KG-TJ&start=1992&view=chart. Accessed 14 Dec 2022

Yeserkepova IB, Cherednichenko AV, Baekenova MK, Bultekov NU, Lebed LV, Akhmadieva ZhK, Tsareva EG, Satenova EZh, Ermakhanova EM (2014) First bio-year report of the Republic of Kazakhstan. Report. Zhassyl Damu JSC, Nur-Sultan

Gulim Abdi has MSc in Operations Management (University of Birmingham, UK), holds a World Bank Certificate for Socio-Environmental Standards (IFC), GRI Certified Training (2022) and Certificate of Honor of the Ministry of Ecology, Geology and Natural Resources of the Republic of Kazakhstan (2021). She has more than 14 years of experience in professional services. The main experience was received from national companies and international projects (Shell, Chevron) on exploration, development, production of oil ESG consulting services (WB, KPMG, RA ESGQ). She works with sustainability reports in accordance with international standards (GRI, SASB, TCFD), regulatory requirements and best practice for major sectoral companies according to the criteria of rating agencies (Sustainalytics, S&P Global CSA, MSCI). She participated in the development of the National Environmental Portal, development of functions on environmental reporting of GHGs, air emissions.

Nurkhat Zhakiyev is a computer modeler of energy systems, analyst in GHG emissions projections and climate change mitigation. He currently working at Astana IT University. He holds a PhD in physics from the Eurasian National University (Kazakhstan, 2015). He was involved in projects modeling the energy sector of Kazakhstan using ELMOD, TIMES energy systems modelling tools at Nazarbayev University, participated in the project "Development of Kazakhstan's National Communication to the UNFCCC and Biennial Report" (In 2015–2021). He was responsible for provision of information about Kazakhstan's activities to mitigate climate change and update of forecasts for 2030 and 2050 in industrial sector. In 2019–2020, he participated in a soft linking project of CGE (top-down) model of Kazakhstan with TIMES-Kazakhstan (bottom-up) energy system model (PMR/World Bank Project for Kazakhstan). In connection with this project, he took part in development of initial data sets, optimization model of energy system of Kazakhstan, in analysis of the various policies and measures to reduce the impact on climate, in analysis of the role of the emissions trading system and development different scenarios. He is a Principal investigator of the project from the Ministry of Science and Higher Education of Kazakhstan "Hybrid modeling of the energy system for development of renewable energy roadmap for Kazakhstan" for 2021–2023.

Shynar Toilybayeva is the Director of the CAREC branch in Kazakhstan, MBA of University of Gloucestershire, UK, MSc of Economics in Management, Kazakh-German University. She coordinated environmental projects of CAREC via green growth (GG) and green economy (GE) approaches in Kazakhstan and Central Asian countries, as well as implementation of the Concept of transition to a low-carbon development. Managed the mainstreaming ecosystem services into country's sectoral and macro economical strategies and policies in the Republic of Kazakhstan. More than ten years of experience in international organizations in the environment development policy, trade and investment, and management.

Open Access This chapter is licensed under the terms of the Creative Commons Attribution 4.0 International License (http://creativecommons.org/licenses/by/4.0/), which permits use, sharing, adaptation, distribution and reproduction in any medium or format, as long as you give appropriate credit to the original author(s) and the source, provide a link to the Creative Commons license and indicate if changes were made.

The images or other third party material in this chapter are included in the chapter's Creative Commons license, unless indicated otherwise in a credit line to the material. If material is not included in the chapter's Creative Commons license and your intended use is not permitted by statutory regulation or exceeds the permitted use, you will need to obtain permission directly from the copyright holder.

Energy Transition in Central Asia

Energy Transition in Central Asia: A Systematic Literature Review

Burulcha Sulaimanova, Indra Overland, Rahat Sabyrbekov, and Roman Vakulchuk

Abstract While there is abundant research on the expansion of renewable energy in developed countries, little attention has been paid to the decarbonisation of energy systems in Central Asia, despite the region's vulnerability to climate change, its rapidly growing domestic energy demand and the abundance of natural resources essential for the energy transition. Based on a systematic review of the literature, this chapter provides a comprehensive overview of the profile and trajectory of research on energy in Central Asia between 1991 and 2022. It finds that there was a shift from focusing on fossil fuels to clean energy around 2019–2020. However, despite recent growth, research on renewables and their significance in Central Asia is still sparse. This review indicates that while American and European researchers took the lead in this field in 2012, China, Japan, Kazakhstan and Russia have emerged as the leading contributors since 2016.

Keywords Central Asia · Clean energy transition · Renewable energy · Systematic literature review · Research networks · Bibliometric visualisation

---

B. Sulaimanova (✉) · R. Sabyrbekov
Organization for Security and Cooperation in Europe (OSCE) Academy, Bishkek, Kyrgyzstan
e-mail: b.sulaimanova@osce-academy.net

R. Sabyrbekov
e-mail: r.sabyrbekov@osce-academy.net

I. Overland · R. Vakulchuk
Norwegian Institute of International Affairs (NUPI), Oslo, Norway
e-mail: ino@nupi.no

R. Vakulchuk
e-mail: rva@nupi.no

© The Author(s) 2023
R. Sabyrbekov et al. (eds.), *Climate Change in Central Asia*,
SpringerBriefs in Climate Studies,
https://doi.org/10.1007/978-3-031-29831-8_6

1 Introduction

While there has been abundant research on renewable energy in developed countries, less attention has been paid to energy system decarbonisation in less developed regions, though the latter are more vulnerable to the adverse effects of climate change, show rapid growth in domestic energy demand, and are rich in natural resources critical for energy transition (Apfel et al. 2021; Vakulchuk et al. 2020; Overland et al. 2019). One such region is Central Asia, which makes a particularly interesting case study owing to its strategic location and resources (Vakulchuk and Overland 2021). After the collapse of the Soviet Union in 1991, Central Asia came to play an important role in the global economy as a source of fossil fuels, primarily oil and gas. This is also reflected in the literature (Vakulchuk 2016; Vakulchuk et al. 2022). However, given the growing global demand for clean energy and materials critical for the energy transition, the countries of Central Asia may be of increasing interest to researchers of clean energy (Vakulchuk and Overland 2021; Eshchanov et al. 2019a, b, c, d).

The aim of this chapter is to provide a comprehensive overview of the profile and trajectory of energy research in Central Asia, with a particular focus on the social sciences. The existing review articles on Central Asia have tended to focus on specific topics in the scenario and sectoral analysis (Mehta et al. 2021; Kaiser and Pulsipher 2007; Karatayev et al. 2021). None has systematically reviewed the literature on energy transition or renewable energy.

This study presents a systematic review of the energy transition literature published between 1991 and 2022 based on data extracted from the Web of Science (WoS) Core Collection database. Using a combination of descriptive statistical analysis and bibliometric visualisation techniques, we address the following research questions: (1) What sub-topics are covered in research on energy issues in the countries of Central Asia? (2) Can a shift from research on fossil fuels to clean energy be identified? (3) What are the most influential articles and research organisations in the field of energy research in Central Asia?

In the next section, we present our strategy for selecting publications, the data extraction and screening processes and visualisation methods. This is followed by our results section, which addresses our research questions, illustrating the empirical results with data visualisations. In the final section, we present our conclusions.

2 Methodology

We based our identification and extraction of publications for review on the Preferred Reporting Items for Systematic Reviews and Meta-Analysis (PRISMA) statement (Page et al. 2021). Our search covered social science publications written in English and listed in the Web of Science (WoS) Social Science Research indexes (SSCI, CPCI-SSH, BKCI-SSH) for the period from 1991 to 2022. Relevant publications were first identified through the WoS Core Collection database by applying the following Boolean search string to the title, abstract and keyword fields:

((TS=(('renewable energy' OR 'solar energy' OR 'solar power' OR 'wind energy' OR 'wind power' OR 'hydropower' OR 'hydroelectric energy' OR 'biofuel' OR 'geothermal energy' OR 'geothermal power' OR 'power plant' OR 'agrofuel' OR 'bioenergy' OR 'green energy' OR 'clean energy' OR 'energy efficiency' OR 'energy consumption' OR 'energy transition' OR 'nuclear energy' OR 'fossil fuel' OR 'coal' OR 'decarbonisation' OR 'low carbon' OR 'petrol*' OR 'gasoline' OR 'oil' OR 'fuel' OR 'electricity' OR 'natural gas') AND ('Central Asia' OR 'Kazakh*' OR 'Kyrgyz*' OR 'Tajik*' OR 'Turkmen*' OR 'Uzbek*'))) AND PY=(1991-2022)) AND LA=(English)

A total of 542 publications were initially retrieved. This was subsequently narrowed down to 305 publications following a screening process including manual relevance checks and the limiting of articles to those published in peer-reviewed journals. The extracted WoS data included the article title, author(s), year of publication, name/country of the research organisation, keywords, citation information and list of references.

The final sample of 305 articles was exported in tab-delimited format to the VOSviewer software. VOSviewer is an application for creating maps and visualisations based on network data (Van Eck and Waltman 2022). It can identify networks among scientific publications, journals, authors, research organisations, countries and/or keywords. Objects of interest (e.g. articles) can be linked, for instance, by co-authorship, citations and bibliographic links. VOSviewer provides three different visualisation modalities: network, overlay and density. In this study, we use overlay figures. An overlay figure represents items with labels and bubbles (Van Eck and Waltman 2022). The greater the weight of the object, the larger the label and the bubble. The colour of an item is determined by its grade, in our case, the grade is the year of publication. By default, the colours range from blue (earliest years) through green to yellow (latest years) (Van Eck and Waltman 2022). See Fig. 1 for an example of an overlay figure.

One of the main limitations of this study is that it is empirically limited to the WoS database. In future research, it would be possible to cover a wider range of databases, though we do not expect that the results would differ substantially. Another limitation is that our dataset includes only papers published in English, while Russian has remained an important language for academic articles on Central Asia and there is also a growing number of publications on the region in Chinese.

3 Results

3.1 Key Trends in Energy Research in Central Asia (Research Questions 1 and 2)

We analysed the co-occurrence of keywords to assess the attention given to different topics and how priorities have evolved. The visualisation in Fig. 1 is based on the

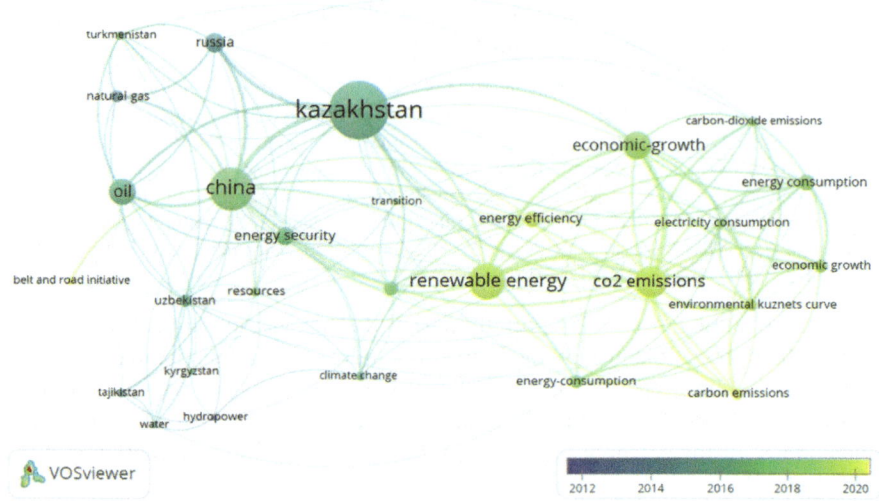

Fig. 1 Main topics in research on energy-related issues in the Central Asian countries (Overlay visualisation of co-occurrence of keywords using VOSviewer)

number of articles in which the same keywords occur. The rationale for this is that the topics that these words represent should be closely related and can illustrate research trends and patterns (Maier et al. 2020) in the popularity of topics over time. Figure 1 uses an overlay visualisation to represent 28 keywords that appear at least 6 times in our 305 articles (out of a total of 1296 keywords used in our analysis). The size of the circles indicates the total number of occurrences of a keyword, while the colour indicates the period when most of the occurrences were noted. VOSviewer automatically assigns a range of years to the period during which most articles were published. The colour of keywords of articles published before 2012 or after 2020 are converted to the 2012 or 2020 colour value, respectively. The distance between the words also indicates their co-occurrence within articles.

The colour coding in Fig. 1 indicates that there has been a shift from fossil fuel-related research to research on the clean energy transition. Thus, keywords such as 'renewable energy' occurred in 32 articles, 'CO_2 emissions' in 28 articles, 'environmental Kuznets curve' in 11. All appear as yellow-green circles, indicating that they have been used more frequently in recent years. By contrast, the keywords 'natural gas' (11) and 'oil' (23) were used more often around 2014–2016, as indicated by the blue/green colours. Analysing the proximity of keywords to one another, we find that the more recent research on clean energy relates more to 'energy efficiency' (14 times), 'energy consumption' (14 times) and 'economic growth' (12 times), as indicated by their size and proximity to each other. By contrast, the keyword 'hydropower' is located far from these other keywords and is coloured purple, indicating that the more recent clean energy research focuses less on hydropower than on other topics. The figure also indicates a lack of publications on 'solar', 'wind energy' and 'bioenergy', as these do not feature in the figure.

Figure 1 indicates that Kazakhstan is the most studied country in the region from an energy perspective. Analysing the distance between the keywords, we can see that the word 'Kazakhstan' is located closer to the clean energy keywords than the other countries. This indicates that Kazakhstan is also the country in which renewable energy topics have been covered most in recent years. By contrast, other Central Asian countries are located at a greater distance from the 'recently occurring' keywords related to clean energy. The distance between keywords indicates that the main topics researched in relation to 'Kyrgyzstan' (6) and 'Tajikistan' (8) are 'hydropower' (6) and 'water' (6), while the colour indicates that these were published in the earlier period. For 'Turkmenistan' (8), the most commonly occurring keywords are 'natural gas' and 'oil', while for 'Uzbekistan' (11) it is 'oil', with colour also indicating earlier publication dates.

To identify key research clusters and hot topics, we analysed bibliographic links between articles with at least 20 citations. This resulted in 6 thematic clusters involving 60 articles (Table 1). Two articles are bibliographically coupled when they link to a common third work in their bibliography. The identified clusters are (A) Energy research from a macroeconomic perspective; (B) Energy security in Central Asia; (C) Natural resources of Kazakhstan; (D) Renewable energy and energy efficiency; (E) Energy geopolitics; and (F) Energy policy (Table 1).

According to the cluster analysis, energy research in Central Asia is most actively researched from the perspective of macroeconomics (Cluster A) and energy security (Cluster B). Cluster A covers the impact of natural resources, energy consumption and carbon dioxide on economic growth and trade (twelve articles in total); the energy-financial development nexus (three articles); and the environment (four articles, including Ansari et al. [2020], which revisit the environmental Kuznets curve). Most of these studies use the statistical method of panel data analysis and focus on multiple countries, both within and outside Central Asia. In Cluster B (energy security), the authors analyse political perspectives on energy corridors, natural gas resources, gas production and pipelines (six articles), and electricity reforms and energy security (eight articles).

3.2 Influential Articles, Authors and Research Organisations (Research Question 3)

Following Maier et al. (2020), we identified the most influential articles from our selection based on the number of citations (Table 2). Among the most cited articles, the three main research topics were the macroeconomic impacts of energy use; energy geopolitics; and climate change mitigation. The most cited article (Sorg et al. 2012) focuses on the impact of climate change on glaciation, the impact of runoff on water availability, and the implications for irrigation, industry and hydropower.

Table 1 Central Asian energy research clusters

Cluster A	Energy from a macroeconomic perspective	Cluster C	Natural resources of Kazakhstan
	Al-Mulali et al. (2012), Ansari et al. (2020), Bildirici and Kayıkçı (2013), Billmeier and Massa (2009), Buehn and Farzanegan (2013), Hafeez et al. (2019), Hasanov et al. (2017), Hasanov et al. (2019), Khan et al. (2014), Mukhtarov et al. (2020), Murshed (2020), Omri et al (2014), Qureshi et al. (2016); Rasoulinezhad and Saboori (2018), Sarkodie et al. (2019), Sun et al. (2019), Wang et al. (2020), Yildirim et al. (2020), and Zhang (2019)		Atakhanova and Howie (2007), Domjan and Stone (2010), Egert and Leonard (2008), Franke et al. (2009), Ipek (2007), Kaiser and Pulsipher (2007), Kalyuzhnova and Nygaard (2008), Lai (2007), Lee (2005), Pomfret (2005), and Zhao (2008)
Cluster B	Energy security in Central Asia	Cluster D	Renewable energy and energy efficiency
	Bilgin (2007), Bilgin (2009), Ericson (2009), Esen and Oral (2016), Knox-Hayes et al. (2013), Le et al (2019), Li et al. (2019), Mohsin et al. (2021), Sarbassov et al. (2013), Sovacool et al. (2012), Sovacool (2016), Taghizadeh-Hesary et al. (2019), Tavana et al. (2012), Wegerich et al. (2007), and Zhang et al. (2019)		Ahmad et al. (2017), Akhanova et al. (2020), Duan et al. (2018), Karatayev et al. (2016), Mahmood and Orazalin (2017), Przychodzen and Przychodzen (2020), Zhang and Bai (2020), and Zhao et al. (2018)
Cluster F	Energy policy	Cluster E	Energy geopolitics
	Koch and Perreault (2019) and Ngoasong (2014)		Chung (2004), Collins and Kearins (2010), Granit et al. (2012), Xunpeng et al. (2017), and Weinthal and Luong (2001)

Figure 2 shows a significant increase in the number of peer-reviewed journal articles and citations between 1991 and 2021. The annual number of citations grew rapidly, reaching 1598 in 2021. The first article in our selection was published in 1992, and until 2007 only a few articles were published annually. The number of peer-reviewed journal articles has been steadily increasing since 2007, accelerating after 2017 and peaking at 41 per year in 2020.

Figure 3 illustrates the number of researchers from their respective countries who have produced energy-related research on Central Asia. The size of a sphere is determined by the number of published articles from that country. The colour of a sphere is determined by the period that most articles were published (blue/purple represents the earlier period, and green/yellow is more recent). Figure 3 depicts 33 countries whose researchers have produced at least three energy articles (of the total of 63

Table 2 The 10 most cited articles

	Article	Author (year)	Journal	Total citations
1	Climate change impacts on glaciers and runoff in Tien Shan (Central Asia)	Sorg et al. (2012)	Nature Climate Change	524
2	Causal interactions between CO_2 emissions, FDI, and economic growth: Evidence from dynamic simultaneous-equation models	Omri et al. (2014)	Economic Modelling	267
3	Exploring the bi-directional long-run relationship between urbanisation, energy consumption, and carbon dioxide emission	Al-Mulali et al. (2012)	Energy	161
4	Energy investment risk assessment for nations along China's Belt & Road Initiative	Duan et al. (2018)	Journal of cleaner production	119
5	Geopolitics of European natural gas demand: Supplies from Russia, Caspian and the Middle East	Bilgin (2009)	Energy Policy	102
6	Global estimates of energy consumption and greenhouse gas emissions	Khan et al. (2014)	Renewable and Sustainable Energy Reviews	88
7	Nexus between energy efficiency and electricity reforms: A DEA-Based way forward for clean power development	Mohsin et al. (2021)	Energy Policy	86
8	Energy crisis, greenhouse gas emissions and sectoral growth reforms: Repairing the fabricated mosaic	Qureshi et al. (2016)	Journal of Cleaner Production	82

(continued)

Table 2 (continued)

	Article	Author (year)	Journal	Total citations
9	Panel estimation for renewable and non-renewable energy consumption, economic growth, CO_2 emissions, the composite trade intensity, and financial openness of the commonwealth of independent states	Rasoulinezhad and Saboori (2018)	Environmental Science and Pollution Research	76
10	Kazakhstan and Azerbaijan as Post-Soviet Rentier States: Resource Incomes and Autocracy as a Double 'Curse' in Post-Soviet Regimes	Franke et al. (2009)	Europe-Asia Studies	68

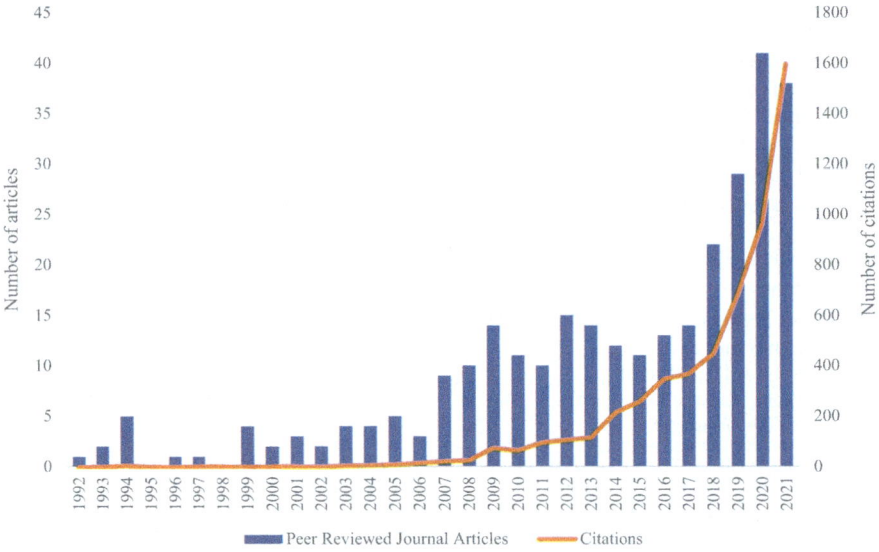

Fig. 2 Publications and the number of citations by year over the 1991–2021 period

countries whose researchers produced our 305 articles). The top contributors are from the US (79 articles), China (57), the UK (37) and Kazakhstan (34). However, there has been a clear dominance in recent years of China, Japan, Kazakhstan and Russia, while researchers from the USA, UK, The Netherlands, Germany and Turkey

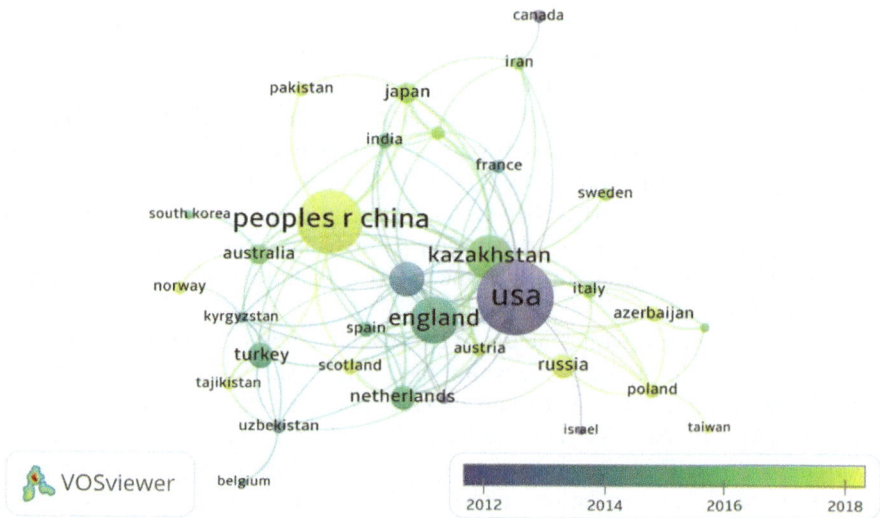

Fig. 3 Countries whose researchers publish the most research on energy in Central Asia (Overlay visualisation of co-authorship link between countries)

were more active between 2012 and 2016 (Fig. 3). This reflects a decline in international scholarly interest in the topic of petroleum resources in Central Asia from 2010 onwards.

The data also showed collaboration between the 33 research organisations that published at least three articles included in our selection (out of the 487 organisations that were involved in the 305 articles). The colour of the circles is determined by the year the articles were published (the most recent being yellow), with proximity indicating tighter collaboration. The greater the number of articles published by an organisation with a collaborating institution, the larger the circle. Thus, greater collaboration in recent years had been seen between Kazakhstan's Nazarbayev University (10 articles and 143 citations), the Asian Development Bank Institute (Japan) (3 articles and 31 citations), the CAREC Institute (China) (3 articles and 7 citations), and Al-Farabi Kazakh National University (6 articles and 31 citations).

4 Discussion and Conclusions

The main objective of our study was to provide a comprehensive overview of trends in energy research in Central Asia. We carried out a systematic literature review and examined 305 articles published between 1991 and 2022 in journals indexed in the Social Science Research indexes, namely SSCI, CPCI-SSH and BKCI-SSH, of the Web of Science Core Collection.

Our bibliometric analysis revealed a trend in research publication topics moving away from fossil fuels towards clean energy in Central Asia between 1991 and 2022. Despite recent growth in the number of articles published on renewable energy in these countries, there is a lack of articles on solar, wind and bioenergy. The top contributors to the literature in recent years are from China, Japan, Kazakhstan and Russia, while researchers from the USA, UK, The Netherlands, Germany and Turkey were more active in the 2012–2016 period. The strongest collaboration between research organisations is observed between Kazakhstan's Nazarbayev University, the Asian Development Bank Institute, the CAREC Institute and Al-Farabi Kazakh National University.

This data provides us with a picture of the academic research networks currently operating in the field of energy transition in Central Asia. Such analysis can help those who are seeking research partners or a research base, those who might want to fund productive collaboration, and those wishing to better understand the dynamics of current research. This study contributes to understanding the clean energy transition in Central Asia; however, for deeper analysis, we recommend exploring more databases and including Russian-language publications.

References

Ahmad S, Nadeem A, Akhanova G, Houghton T, Muhammad-Sukki F (2017) Multi-criteria evaluation of renewable and nuclear resources for electricity generation in Kazakhstan. Energy 141:1880–1891

Akhanova G, Nadeem A, Kim JR, Azhar S (2020) A multi-criteria decision-making framework for building sustainability assessment in Kazakhstan. Sustain Cities Soc 52:101842

Al-Mulali U, Sab CNBC, Fereidouni HG (2012) Exploring the bi-directional long run relationship between urbanization, energy consumption, and carbon dioxide emission. Energy 46(1):156–167

Ansari MA, Haider S, Khan NA (2020) Environmental Kuznets curve revisited: an analysis using ecological and material footprint. Ecol Ind 115:106416

Apfel D, Haag S, Herbes C (2021) Research agendas on renewable energies in the Global South: a systematic literature review. Renew Sustain Energy Rev 148:111228

Atakhanova Z, Howie P (2007) Electricity demand in Kazakhstan. Energy Policy 35(7):3729–3743

Bildirici ME, Kayıkçı F (2013) Effects of oil production on economic growth in Eurasian countries: panel ARDL approach. Energy 49:156–161

Bilgin M (2007) New prospects in the political economy of inner-Caspian hydrocarbons and western energy corridor through Turkey. Energy Policy 35(12):6383–6394

Bilgin M (2009) Geopolitics of European natural gas demand: supplies from Russia, Caspian and the Middle East. Energy Policy 37(11):4482–4492

Billmeier A, Massa I (2009) What drives stock market development in emerging markets—institutions, remittances, or natural resources? Emerg Mark Rev 10(1):23–35

Buehn A, Farzanegan MR (2013) Hold your breath: a new index of air pollution. Energy Econ 37:104–113

Chung CP (2004) The Shanghai Co-operation Organization: China's changing influence in Central Asia. China Q 180:989–1009

Collins EM, Kearins K (2010) Delivering on sustainability's global and local orientation. Acad Manag Learn Educ 9(3):499–506

Domjan P, Stone M (2010) A comparative study of resource nationalism in Russia and Kazakhstan 2004–2008. Eur Asia Stud 62(1):35–62

Duan F, Ji Q, Liu BY, Fan Y (2018) Energy investment risk assessment for nations along China's Belt & Road Initiative. J Clean Prod 170:535–547

Egert B, Leonard CS (2008) Dutch disease scare in Kazakhstan: is it real? Open Econ Rev 19(2):147–165

Ericson RE (2009) Eurasian natural gas pipelines: the political economy of network interdependence. Eurasian Geogr Econ 50(1):28–57

Esen V, Oral B (2016) Natural gas reserve/production ratio in Russia, Iran, Qatar and Turkmenistan: a political and economic perspective. Energy Policy 93:101–109

Eshchanov B, Abylkasymova A, Aminjonov F, Moldokanov D, Overland I, Vakulchuk R (2019a) Renewable energy policies of the Central Asian countries. Cent Asia Reg Data Rev 16:1–4

Eshchanov B, Abylkasymova A, Aminjonov F, Moldokanov D, Overland I, Vakulchuk R (2019b) Wind power potential of the Central Asian countries. Cent Asia Reg Data Rev 17:1–7

Eshchanov B, Abylkasymova A, Aminjonov F, Moldokanov D, Overland I, Vakulchuk R (2019c) Solar power potential of the Central Asian countries. Cent Asia Reg Data Rev 18:1–7

Eshchanov B, Abylkasymova A, Aminjonov F, Moldokanov D, Overland I, Vakulchuk R (2019d) Hydropower potential of the Central Asian countries. Cent Asia Reg Data Rev 19:1–7

Franke A, Gawrich A, Alakbarov G (2009) Kazakhstan and Azerbaijan as post-Soviet rentier states: resource incomes and autocracy as a double 'curse' in post-Soviet regimes. Eur Asia Stud 61(1):109–140

Granit J, Jägerskog A, Lindström A, Björklund G, Bullock A, Löfgren R, De Gooijer G, Pettigrew S (2012) Regional options for addressing the water, energy and food nexus in Central Asia and the Aral Sea basin. Int J Water Resour Dev 28(3):419–432

Hafeez M, Yuan C, Khelfaoui I, Sultan Musaad OA, Waqas Akbar M, Jie L (2019) Evaluating the energy consumption inequalities in the one belt and one road region: implications for the environment. Energies 12(7):1358

Hasanov F, Bulut C, Suleymanov E (2017) Review of energy-growth nexus: a panel analysis for ten Eurasian oil exporting countries. Renew Sustain Energy Rev 73:369–386

Hasanov FJ, Mikayilov JI, Mukhtarov S, Suleymanov E (2019) Does CO_2 emissions–economic growth relationship reveal EKC in developing countries? Evidence from Kazakhstan. Environ Sci Pollut Res 26(29):30229–30241

İpek P (2007) The role of oil and gas in Kazakhstan's foreign policy: looking east or west? Eur Asia Stud 59(7):1179–1199

Kaiser MJ, Pulsipher AG (2007) A review of the oil and gas sector in Kazakhstan. Energy Policy 35(2):1300–1314

Kalyuzhnova Y, Nygaard C (2008) State governance evolution in resource-rich transition economies: An application to Russia and Kazakhstan. Energy Policy 36(6):1829–1842

Karatayev M, Hall S, Kalyuzhnova Y, Clarke ML (2016) Renewable energy technology uptake in Kazakhstan: policy drivers and barriers in a transitional economy. Renew Sustain Energy Rev 66:120–136

Karatayev M, Lisiakiewicz R, Gródek-Szostak Z, Kotulewicz-Wisińska K, Nizamova M (2021) The promotion of renewable energy technologies in the former Soviet bloc: why, how, and with what prospects? Energy Rep 7:6983–6994

Khan MA, Khan MZ, Zaman K, Naz L (2014) Global estimates of energy consumption and greenhouse gas emissions. Renew Sustain Energy Rev 29:336–344

Knox-Hayes J, Brown MA, Sovacool BK, Wang Y (2013) Understanding attitudes toward energy security: results of a cross-national survey. Glob Environ Chang 23(3):609–622

Koch N, Perreault T (2019) Resource nationalism. Prog Hum Geogr 43(4):611–631

Lai HH (2007) China's oil diplomacy: is it a global security threat? Third World Q 28(3):519–537

Le TH, Chang Y, Taghizadeh-Hesary F, Yoshino N (2019) Energy insecurity in Asia: a multi-dimensional analysis. Econ Model 83:84–95

Lee PK (2005) China's quest for oil security: oil (wars) in the pipeline? Pac Rev 18(2):265–301

Li JX, Chen YN, Xu CC, Li Z (2019) Evaluation and analysis of ecological security in arid areas of Central Asia based on the emergy ecological footprint (EEF) model. J Clean Prod 235:664–677

Mahmood M, Orazalin N (2017) Green governance and sustainability reporting in Kazakhstan's oil, gas, and mining sector: evidence from a former USSR emerging economy. J Clean Prod 164:389–397

Maier D, Maier A, Așchilean I, Anastasiu L, Gavriș O (2020) The relationship between innovation and sustainability: a bibliometric review of the literature. Sustainability 12(10):4083

Mehta K, Ehrenwirth M, Trinkl C, Zörner W, Greenough R (2021) The energy situation in Central Asia: a comprehensive energy review focusing on rural areas. Energies 14(10):2805

Mohsin M, Hanif I, Taghizadeh-Hesary F, Abbas Q, Iqbal W (2021) Nexus between energy efficiency and electricity reforms: a DEA-based way forward for clean power development. Energy Policy 149:112052

Mukhtarov S, Humbatova S, Seyfullayev I, Kalbiyev Y (2020) The effect of financial development on energy consumption in the case of Kazakhstan. J Appl Econ 23(1):75–88

Murshed M (2020) Are Trade Liberalization policies aligned with renewable energy transition in low and middle income countries? An instrumental variable approach. Renewable Energy 151:1110–1123

Ngoasong MZ (2014) How international oil and gas companies respond to local content policies in petroleum-producing developing countries: a narrative enquiry. Energy Policy 73:471–479

Omri A, Nguyen DK, Rault C (2014) Causal interactions between CO_2 emissions, FDI, and economic growth: Evidence from dynamic simultaneous-equation models. Econ Model 42:382–389

Overland I, Bazilian M, Uulu TI, Vakulchuk R, Westphal K (2019) The GeGaLo index: geopolitical gains and losses after energy transition. Energ Strat Rev 26:100406

Page MJ, McKenzie JE, Bossuyt PM, Boutron I, Hoffmann TC, Mulrow CD et al (2021) The PRISMA 2020 statement: an updated guideline for reporting systematic reviews. BMJ 372:n71. https://doi.org/10.1136/bmj.n71

Pomfret R (2005) Kazakhstan's economy since independence: does the oil boom offer a second chance for sustainable development? Eur Asia Stud 57(6):859–876

Przychodzen W, Przychodzen J (2020) Determinants of renewable energy production in transition economies: a panel data approach. Energy 191:116583

Qureshi MI, Rasli AM, Zaman K (2016) Energy crisis, greenhouse gas emissions and sectoral growth reforms: repairing the fabricated mosaic. J Clean Prod 112:3657–3666

Rasoulinezhad E, Saboori B (2018) Panel estimation for renewable and non-renewable energy consumption, economic growth, CO_2 emissions, the composite trade intensity, and financial openness of the commonwealth of independent states. Environ Sci Pollut Res 25(18):17354–17370

Sarbassov Y, Kerimray A, Tokmurzin D, Tosato G, De Miglio R (2013) Electricity and heating system in Kazakhstan: exploring energy efficiency improvement paths. Energy Policy 60:431–444

Sarkodie SA, Strezov V, Jiang Y, Evans T (2019) Proximate determinants of particulate matter ($PM_{2.5}$) emission, mortality and life expectancy in Europe, Central Asia, Australia, Canada and the US. Sci Total Environ 683:489–497

Sorg A, Bolch T, Stoffel M, Solomina O, Beniston M (2012) Climate change impacts on glaciers and runoff in Tien Shan (Central Asia). Nat Clim Chang 2(10):725–731

Sovacool BK (2016) Differing cultures of energy security: an international comparison of public perceptions. Renew Sustain Energy Rev 55:811–822

Sovacool BK, Valentine SV, Bambawale MJ, Brown MA, de Fátima Cardoso T, Nurbek S, Suleimenova G, Li J, Xu Y, Jain A, Alhajji AF, Zubiri A (2012) Exploring propositions about perceptions of energy security: an international survey. Environ Sci Policy 16:44–64

Sun H, Attuquaye Clottey S, Geng Y, Fang K, Clifford Kofi Amissah J (2019) Trade openness and carbon emissions: evidence from belt and road countries. Sustainability 11(9):2682

Taghizadeh-Hesary F, Yoshino N, Rasoulinezhad E, Chang Y (2019) Trade linkages and transmission of oil price fluctuations. Energy Policy 133:110872

Tavana M, Pirdashti M, Kennedy DT, Belaud JP, Behzadian M (2012) A hybrid Delphi-SWOT paradigm for oil and gas pipeline strategic planning in Caspian Sea basin. Energy Policy 40:345–360

Vakulchuk R (2016) Public administration reform and its implications for foreign petroleum companies in Kazakhstan. Int J Public Adm 39(14):1180–1194

Vakulchuk R, Overland I (2021) Central Asia is a missing link in analyses of critical materials for the global clean energy transition. One Earth 4(12):1678–1692

Vakulchuk R, Overland I, Scholten D (2020) Renewable energy and geopolitics: a review. Renew Sustain Energy Rev 122:109547

Vakulchuk R, Isataeva A, Kolodzinskaia G, Overland I, Sabyrbekov R (2022) Fossil fuels in Central Asia: trends and energy transition risks. Cent Asia Reg Data Rev 28:1–6

Van Eck NJ, Waltman L (2022) VOSviewer manual. Manual for VOSviewer version 1.6.18. Univeristeit Leiden, Leiden

Wang Q, Lin J, Zhou K, Fan J, Kwan MP (2020) Does urbanization lead to less residential energy consumption? A comparative study of 136 countries. Energy 202:117765

Wegerich K, Olsson O, Froebrich J (2007) Reliving the past in a changed environment: hydropower ambitions, opportunities and constraints in Tajikistan. Energy Policy 35(7):3815–3825

Weinthal E, Luong PJ (2001) Energy wealth and tax reform in Russia and Kazakhstan. Resour Policy 27(4):215–223

Xunpeng S, Variam HMP, Tao J (2017) Global impact of uncertainties in China's gas market. Energy Policy 104:382–394

Yıldırım S, Gedikli A, Erdoğan S, Yıldırım DÇ (2020) Natural resources rents-financial development nexus: evidence from sixteen developing countries. Resour Policy 68:101705

Zhang J (2019) Oil and gas trade between China and countries and regions along the 'Belt and Road': a panoramic perspective. Energy Policy 129:1111–1120

Zhang L, Bai W (2020) Risk assessment of China's natural gas importation: a supply chain perspective. Sage Open 10(3):2158244020939912

Zhang X, Zhang H, Yuan J (2019) Economic growth, energy consumption, and carbon emission nexus: fresh evidence from developing countries. Environ Sci Pollut Res 26(25):26367–26380

Zhang YJ, Jin YL, Shen B (2020) Measuring the energy saving and CO_2 emissions reduction potential under China's belt and road initiative. Comput Econ 55(4):1095–1116

Zhao S (2008) China's global search for energy security: cooperation and competition in Asia-Pacific. J Contemp China 17(55):207–227

Zhao C, Zhang H, Zeng Y, Li F, Liu Y, Qin C, Yuan J (2018) Total-factor energy efficiency in BRI countries: an estimation based on three-stage DEA model. Sustainability 10(1):278

Burulcha Sulaimanova is a Postdoctoral Research Fellow at the OSCE Academy, specializing in energy economics, labor economics, and applied economic analysis in the Central Asian countries. She holds a PhD in Economics from Kyrgyz-Turkish Manas University in Kyrgyzstan and previously worked as an Assistant Professor at the same university. Her research focuses on the economic challenges and opportunities faced by Central Asian countries, including energy security, labor market dynamics, and sustainable development.

Indra Overland is Research Professor and Head of the Research Group on Climate and Energy at the Norwegian Institute of International Affairs (NUPI). He previously headed the Russia and Eurasia Research Group at NUPI and has worked on Central Asia since 2001. He completed his PhD at the University of Cambridge, followed by a three-year post-doctoral project on Central Asia and the South Caucasus. He has carried out fieldwork in all Central Asian states and has been responsible for cooperation between the OSCE Academy in Kyrgyzstan and NUPI since 2007. Every year he teaches MA students from all Central Asian countries on energy issues and hosts

4-5 Central Asian students in Norway. He is (co)author of *Caspian Energy Politics* (Routledge), *China's Belt and Road Initiative Through the Lens of Central Asia* (Routledge), *Kazakhstan: Civil Society and Natural Resource Policy in Kazakhstan* (Palgrave) and *Renewable Energy Policies of the Central Asian Countries* (CADGAT). He is a contributing author to the IPCC's Sixth Assessment Report (Working Group II).

Rahat Sabyrbekov is an environmental economist who specializes in decarbonization, climate change and energy transition. Rahat is a Postdoctoral Fellow at OSCE Academy and Visiting Fellow at Davis Center at Harvard University. He obtained his PhD from School of Economics and Business at Norwegian University of Life Sciences. He received his Master's degree from University of Birmingham, the United Kingdom. He teaches Economics of Sustainable Management of Mineral Resources course at the OSCE Academy. His recent publications include *Putting the Foot Down: Accelerating EV Uptake in Kyrgyzstan* (2023), *Know Your Opponent: Which Countries might Fight the European Carbon Border Adjustment Mechanism?* (2022), *Fossil Fuels in Central Asia: Trends and Energy Transition Risks* (2022).

Roman Vakulchuk is Senior Researcher at the Norwegian Institute of International Affairs (NUPI) in Oslo and holds a PhD degree in Economics from Jacobs University Bremen in Germany. He specializes in Central Asia and Southeast Asia and his main research interests are economic transition, trade, energy, climate change and investment policy. Vakulchuk has served as project leader in research projects organized by the Asian Development Bank (ADB), the World Bank, the Global Development Network (GDN), the Natural Resource Governance Institute (NRGI) and others. In 2018, he worked as governance expert for OECD's mission in Kazakhstan and advised the government on privatization reform. In 2013, Vakulchuk was awarded the Gabriel Al-Salem International Award for Excellence in Consulting. Recent publications include *Seizing the Momentum. EU Green Energy Diplomacy Towards Kazakhstan* (2021), *Discovering Opportunities in the Pandemic? Four Economic Response Scenarios for Central Asia* (2020) and *Renewable Energy and Geopolitics: A Review* (2020).

Open Access This chapter is licensed under the terms of the Creative Commons Attribution 4.0 International License (http://creativecommons.org/licenses/by/4.0/), which permits use, sharing, adaptation, distribution and reproduction in any medium or format, as long as you give appropriate credit to the original author(s) and the source, provide a link to the Creative Commons license and indicate if changes were made.

The images or other third party material in this chapter are included in the chapter's Creative Commons license, unless indicated otherwise in a credit line to the material. If material is not included in the chapter's Creative Commons license and your intended use is not permitted by statutory regulation or exceeds the permitted use, you will need to obtain permission directly from the copyright holder.

A 'Steppe' into the Void: Central Asia in the Post-oil World

Morena Skalamera

Abstract Kazakhstan, Turkmenistan and Uzbekistan are petrostates and therefore trapped by the global energy transition. This chapter delves into the nexus between the effects of the energy transition, international stability and regime stability in Central Asia's fossil-fuel dominated economies—a nexus of increasing theoretical and policy relevance as we enter a post-oil era in world politics. The Central Asian hydrocarbon producers are torn between their own aspirations to shift to a low-carbon economy and the vested interests of their elites, which are embedded in fossil-fuel dependency. Despite making international commitments to energy transition and developing policy frameworks to expand the renewable energy sector, the Central Asian petrostates have continued using foreign policy to seek fossil fuel revenue by forging new international trade and investment relations outside of the region. The chapter particularly highlights an under-researched aspect of the global energy transition, namely the role of informal elites in influencing foreign policy strategies, and in undermining energy transitions at the local level in doing so.

Keywords Petrostates · Energy transition · Central Asia · Vested interests · Foreign policy

1 Introduction

This chapter focuses on the ways in which the Central Asian petrostates have altered their foreign policies, domestic efforts and growth strategies in response to the energy transition, and the repercussions of this for domestic and international stability. It dwells on a pair of related issues: regional energy geopolitics and the regime stability of individual Central Asian states. To investigate the links between these, the study

M. Skalamera (✉)
Leiden University, Leiden, The Netherlands
e-mail: m.skalamera@hum.leidenuniv.nl

Belfer Center, Harvard Kennedy School, Cambridge, MA, USA

engages three bodies of knowledge: energy geopolitics and the shift from hydrocarbons to renewable energy (O'Sullivan et al. 2017; Van de Graaf and Bradshaw 2018; Scholten et al. 2020); domestic regime stability in petrostates (Beblawi and Luciani 1987; Acemoglu and Robinson 2006; Morrison 2009); and the political economy of energy in Central Asia (Skalamera 2017; Shadrina 2020; Vakulchuk and Overland 2021). In doing so, this chapter presents a new framework for understanding the relationship between the effects of the global energy transition and the interests of hydrocarbon producers in the under-studied region of Central Asia.

In the literature on revenue loss in resource-based economies, diversification is often presented as a necessary remedy (Shehabi 2019). These studies often overlook petrostates' attempts to forestall change and maintain the status quo through strategic foreign policy shifts. In this chapter I argue that there has been a far-reaching shift in energy relations between the Central Asian petrostates and major fossil fuel importers—the EU and China. The two main drivers of this shift are China's foray into the region as a new energy hegemon and the effects of the energy transition. Regarding the latter, some emphasize the constitutive role of Western climate change thinking in socializing post-Soviet leaders and persuading them—or not—to accelerate their own shift to renewables (e.g. Koch and Tynkkynen 2019). By contrast, I argue that Central Asian leaders' behaviour has been influenced by the lessons they have drawn from observing the consequences of the energy transition on the hydrocarbon portfolios of major energy companies. I argue in particular that the ongoing energy transition has driven *other* relevant hydrocarbon producers (Russia and the United Arab Emirates) to invest in the Central Asian hydrocarbon sector to decrease the pace of regional decarbonisation. Meanwhile, the Central Asian petrostates themselves have adapted in two contrasting ways: by adopting new 'green' identities and policies on the one hand, and by seeking the support of other hydrocarbon producers—and China—for 'old' oil and gas investments and sales on the other.

The chapter first reviews the renewable energy (RE) policies and targets set by the three Central Asian petrostates, Kazakhstan, Uzbekistan and Turkmenistan. Because these RE strategies have been discussed in detail elsewhere (e.g. Eshchanov et al. 2019a, b, c, d), I here examine *to what extent* and *how* these policies have been adopted and implemented. I then explain how national leaders and decision-makers have drawn lessons from recent contemporary shifts in the RE portfolios of European countries in ways that have greatly affected their foreign policies, and ultimately, shaped their decisions to turn to China, Russia, Turkey and the Gulf countries for oil, gas and nuclear trade and investment. Ultimately, I shine a light on a critical, yet under-researched, aspect of the energy transition, namely the role of informal elites in undermining local energy transitions by protecting their own fossil-fuel-dependent interests. I conclude that a form of foreign policy decision-making that is highly volatile and unpredictable because it is susceptible to fossil fuel-related revenues is likely in view of the local elites' dependence on hydrocarbon rents.

2 Renewable Energy Policies and Targets

In response to the global transition to renewable energy, Central Asian states are striving to shake their over-reliance on hydrocarbon-dependent growth. In addition to abundantly available hydropower resources (Eshchanov et al. 2019d), the total amount of solar radiation received by the five Central Asian countries is sufficient to generate 20 times more electricity than they currently generate, (Eshchanov et al. 2019c), while the resource potential for wind power in all the Central Asian countries is even higher (Eshchanov et al. 2019b). Thus far, the high capital costs of RE projects, coupled with a lack of legislative frameworks for the sector, have hindered development. However, RE costs are now falling rapidly and laws are being put in place to foster RE development. Yet, hydrocarbons still dominate domestic electricity generation and the region's wider economies. In Kazakhstan, solar and wind energy are only slowly emerging, while they are entirely new technologies in Uzbekistan and Turkmenistan. Around much of Central Asia, hydropower and oil and gas-fired power plants still account for almost all electricity generation (Dudley 2020).

Between 2019 and 2020, the generation of electricity by renewable sources in Kazakhstan grew by 74% to 3.24 billion kWh (3% of the total energy generated in the country [Energy Central News 2021]). To encourage an attractive investment climate and to establish the framework needed for RE investment, Central Asia is seeing significant structural and regulatory reforms in the electricity sector and wider economies (Wishart and Abidi 2021). Kazakhstan is the first country in Central Asia to pass carbon-pricing legislation and establish a national Emissions Trading Scheme (ETS) (see Abdi et al., this volume).

In addition to current RE use, all countries of the region have precise targets for RE capacity expansion (Eshchanov et al. 2019a).[1] However, the only central Asian state to have developed meaningful amounts of non-hydropower RE to date is Kazakhstan, which has met its target of 3% of the country's energy output from renewable energy sources (RES) by the end of 2020, and aims for RES to generate 10% of its energy needs by 2030 and 50% by 2050. Uzbekistan aims to grow the share of renewables, including hydropower, in its electricity generation to 25% by 2030 (Eurasianet 2021). This suggests that Kazakhstan, and to a lesser degree other Central Asian petrostates, are recognizing the political benefits and opportunities of being at the forefront of the energy transition.

The other Central Asian nations are likely to hasten their own energy transition as incumbents adapt to new conditions or learn from Kazakhstan's experience. Policies that foster renewables have achieved institutional acceptance and successful implementation chiefly when their advocates have been able to link environmental goals with economic ones (Blondeel et al. 2019). However, as I will explore in more detail below, issues of prestige, status and leadership are also critical. The different levels of renewable energy sector development across the countries of Central Asia indicate

---

[1] Turkmenistan became the last of the five Central Asian republics to introduce a renewable energy strategy and pass a law on renewable energy in March 2021 (Wishart and Abidi 2021).

the importance of individual social and political agents in fostering decarbonisation, even as structural changes (a system-wide energy transition) condition their behaviour. Kazakhstan has been more persuasive than its neighbours in framing green ideas in such a way that they resonate with relevant audiences. Thus, as I demonstrate below, the transition to RE is most likely to lead to tangible results in Turkmenistan and Uzbekistan once the old elites are able to discern a reasonably bright future for themselves in the new social order.

3 A Strengthened Transition Profile: The Role of China

As the Central Asian petrostates expand their renewable energy sectors to varying degrees, the role of Chinese business is set to increase, with disruptive effects on Russia's energy policy in the region. Through its Belt and Road Initiative (BRI) and deepening of energy trade and ties with the former Soviet states, China has rapidly expanded its economic presence in Central Asia. Until recently, BRI energy investment was focused on fossil fuels. Yet, in 2020, even as Beijing's overall BRI spending slowed, renewables overtook fossil fuels for the first time in BRI energy investments, increasing from 38 to 57% in one year (Eurasianet 2021). China's investments have long been influenced by its know-how. China now produces 70% of the world's solar panels and holds a leading position in battery manufacturing, along with access to the raw materials needed to make them, and it is also a major wind turbine manufacturer.

China offered Kazakhstan a taste of its technologies, gifting a 1 MW solar plant to the Alatau Innovation Park near Almaty in a 2011 agreement between the two countries, intending to demonstrate Chinese know-how and encourage Kazakhstan to look east for its green energy needs. These overtures paid off, and in June 2018, Ningbo-based Risen Energy began work on a $39 million 40 MW solar photovoltaic plant in Karaganda. The same company is building a 63 MW solar photovoltaic power plant in Chulakkurgan, north of Shymkent (Eurasianet 2020). China is also well represented in Uzbekistan's new renewable projects, but it is not the only player.[2] Firms from Saudi Arabia and the United Arab Emirates (UAE) have also secured contracts to develop solar power projects. In October 2019, the Uzbek Energy Ministry announced that UAE firm Masdar had been awarded a contract to build a 100 MW solar park in Navoi Province.[3]

---

[2] There were reports in September 2019 that Chinese company Lioaning Lide was building a wind farm in the Gijduvan district of Uzbekistan's Bukhara Province.

[3] *Renewables Now*, Masdar wins solar tender in Uzbekistan with bid of USD 26.79/MWh, October 7, 2019; Bruce Pannier, Fossil Fuel Giants Kazakhstan, Uzbekistan Slowly Going Green, RFERL, February 8, 2020.

4 Structural Obstacles to the Growth of the RE Sector

For almost three decades, the abundance of oil, gas and other natural resources in Kazakhstan, Uzbekistan and Turkmenistan has underpinned an economic reliance on raw material exports (Vakulchuk and Overland 2021). Now the oil and gas-rich countries, Kazakhstan *in primis*, are making substantial commitments to a 'green economy' through policies, programmes, action plans and public statements at both international and domestic levels. Diversification, however, is easier said than done, particularly when it comes to promoting stronger exports outside the oil and gas sector, as various political forces create pressure to continue investing in fossil fuels.

Thus, despite signs that Central Asian countries are eager to create low-carbon opportunities, a number of barriers hinder the development of the region's renewable energy sector. These include patchy regulatory frameworks, limited infrastructure, difficulties in attracting finance and technical expertise, insufficient public awareness of the benefits of RE policy, and insufficient data and information to evaluate current efforts and future potential (Laldjebaev et al. 2021). The lack of effective policy, however, appears to be the largest impediment. Most crucially, current reforms have failed to spark significant growth in the renewable electricity sector as the laws that have been adopted suffer due to a lack of actionable measures or failure to allocate responsibility for implementation (Shadrina 2020). Reforms wither due to ineffective implementation by economic elites, especially when self-interested networks perceive that turning points, such as the energy transition, will cut them off from access to power and assets (Skalamera Groce 2020).

Despite a flurry of formal strategies and programmes to increase RE deployment in all three countries, the use of informal institutional designs, practices and procedures for implementation means that rules on 'good environmental governance' are much easier to draw up than to enforce. For instance, in Uzbekistan, despite ambitious goals, the only notable progress appears to be the 100 MW wind farm under construction in Navoi Province, while three out of five solar projects that were announced have now been postponed (Eshchanov et al. 2019b). Efforts to promote solar projects have been undermined by a number of factors, including the fact that the state-owned electricity company is yet to be unbundled, while RE policies lack useful goal benchmarks and comprehensive long-term development strategy (Laldjebaev et al. 2021). Similarly, in Turkmenistan, despite a recently adopted law on RE sector development, there is no strategy for the actual development of RE sources. With no roadmap for implementation, there is also no monitoring or follow-up on RE plans (Laldjebaev et al. 2021).

In Kazakhstan, frequent changes in legislation and the slow translation of policy ideas into action deter international investors, while support for fossil fuels in domestic supply and for export remains strong (Skalamera Groce 2020; Wishart and Abidi 2021). Public awareness of (and support for) the energy transition is notably higher in Kazakhstan than in the other Central Asia petrostates, yet corruption among political and economic elites is increasingly severe (Kudaibergenova and Laruelle 2022). Hydrocarbon trade provides significant material benefits to certain groups of

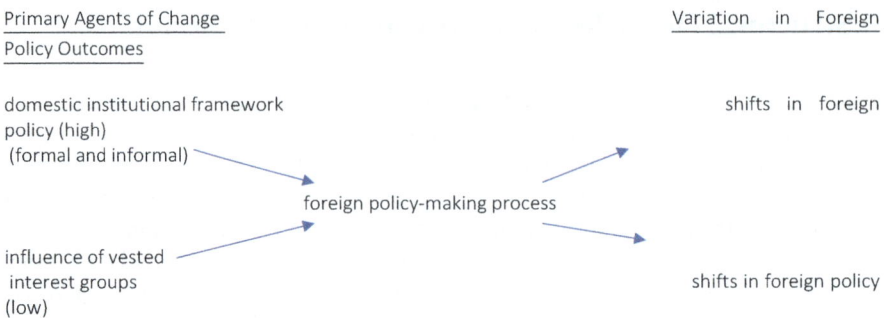

Fig. 1 Foreign policy strategies under conditions of variable hydrocarbon revenues (i.e. energy transition)

people, and actual reform is avoided because of its high opportunity costs, namely, the potential loss of the benefits to those with vested interests (see Fig. 1).

Lucrative oil, gas and raw material rents tend to benefit corrupt leaders and elites, foster corruption, and deepen oppression as leaders and policymakers use them not only to support their own affluent lifestyles but also to expand their political control by subverting opponents and obstructing projects and policies that threaten their source of income (Fattouh et al. 2019). Given the rigidity in government spending, oil policy is 'elevated' to a cure-all to boost revenues and address economic turmoil. Meanwhile, oil rents, and thus pre-existing fiscal buffers, tend to erode over time, leading to a vicious cycle that perpetuates the fiscal deficit (Fattouh 2021).

In other words, declining oil and gas revenues limit the government's capacity to ensure autocratic stability through high social spending commitments and patronage measures targeting the wider population. In January 2022, civil unrest over fuel price hikes in Kazakhstan demonstrated just that; the population had little willingness to accept possible budget cuts and unsubsidized energy prices. This also shows that economic turmoil from plunging hydrocarbon revenues will be disruptive and could lead to political crackdowns and a more volatile foreign policy, as the elites seek to hold on to power (Skalamera, 2020).

5 A Shift in Power Relations

Central Asian state control over a diminishing hydrocarbon market, coupled with the onward global march of the energy transition, may result in a dramatic shift in power relations.[4] To the long-term structural means of entrenching domestic power and the exercise of influence over other countries will be added the impact of 'strategic

---

[4] Even as strategic hydrocarbons are still essential to industrialized countries, especially in the Asia-Pacific, recent technological innovations have increased the role of renewables. Abundance of global oil and gas supply has also reduced petrostates' power.

shocks', such as a Dutch court's finding against Royal Dutch Shell and the subsequent order to slash emissions harder and faster than planned.[5] In this context, the energy transition has resulted in two simultaneous and contradictory responses: a formal response that is broadly supportive of the transition, and an informal response that seeks to preserve individual elite advantage, essentially acting to undermine the transition at the local level. Thus, on the one hand, and despite lagging investment and stalled project timelines, Central Asia is formally embracing the global energy transition (Vakulchuk and Overland 2021), yet plenty of obstacles remain, including weak governance, poor infrastructure and grand corruption that cripple economic development. This brings us to the more informal side of Central Asia's 'adaptation' to the global energy transition.

Informal elites might be in a position to use their influence to switch to alternative buyers of traditional hydrocarbons, looking elsewhere to maintain dependence on the sale of oil, gas and other raw materials, and thus slow the pace of the energy transition. The Central Asian petrostates have focused on strengthening energy, trade and diplomatic ties with other oil and gas-rich countries, such as Russia and Iran, and most significantly, making massive investments into their own untapped hydrocarbon and mineral supplies. This may be a factor contributing to Russia's renewed presence in the region. Oil and gas revenues are then used to build up extensive military-security apparatuses, which are occasionally used for political repression of dissident voices, as we saw in 2021 following the fuel protests in Kazakhstan.

6 Diversifying Hydrocarbon Revenue Sources

While global consumption of all fossil fuels decreased in 2020, with the largest declines in North America (-8.0%) and Europe (-7.8%), the Asia–Pacific region recorded the lowest decrease. China ($+2.1\%$) is the only major country where fossil fuel consumption actually increased in 2020 (BP Statistical Review of World Energy 2021). While many European states have reduced their dependence on oil and gas imports from Central Asia, China has increased its reliance on hydrocarbons from the region. Fossil fuels still account for nearly 84% of global consumption and the Asia–Pacific region alone accounts for nearly 50% of that consumption.[6] This presents Central Asian petrostates with an alternative market opportunity for the next 10–20 years at least.

Sandwiched between regional powers China and Russia, Kazakhstan, for instance, is looking to diversify its portfolio investments into South East Asia. The government is now investing in start-ups across East Asia in an effort to strategically position itself as the region's financial gateway to Europe and the Middle East. China and

---

[5] In May 2021, a civil court in The Netherlands ruled that by 2030 Royal Dutch Shell must cut its CO_2 emissions by 45% compared to 2019 levels (Bloomberg 2021; Reuters 2021).

[6] Statista, Primary energy consumption worldwide from 2010 to 2020 by region, available at: https://www.statista.com/statistics/263457/primary-energy-consumption-by-region/.

Russia, however, still loom large in these efforts to diversify away from Europe. In Kazakhstan, China has a 24% stake in oil production and a 13% stake in gas production. In 2019, it was reported that Kazakhstan would divert some oil from Europe to China and double its gas exports to China (Reuters 2019). Russia has also made itself a useful and enthusiastic ally, not only in delaying the energy transition but in counterbalancing China's heavy energy investments in the region. Kazakhstan and Russia have recently signed an MoU to pursue the construction of energy corridors to China and Europe and to establish joint infrastructure for the sale of compressed natural gas (CNG) (Vestnik Kavkaza 2022). According to the MoU, the energy ministries of Russia and Kazakhstan aspire to promote natural gas as a 'bridge fuel' on the Europe-Western China international transport route.

Russia's burgeoning external influence has focused on spurring joint ventures in hydrocarbon and nuclear energy development. The flow of Russian gas and oil investment to exert control over fuel-dependent Central Asia might now result in a firmer and more combative stance towards the West. As the energy transition advances, Russian companies are developing new and smaller-sized oil and gas fields or purchasing shares from Western companies leaving the region. In October 2021, Kazakhstan's President Tokayev revealed that Russia's oil giant Lukoil would help his country develop Khazar, an oil field located next to the giant Kashagan field in Kazakhstan's portion of the Caspian Sea (Caspian News 2021a). In 2019, Shell scaled back its oil and gas exploration plans for Khazar due to 'the low profitability […] against the background of high capital expenditures', in what was likely a divestment effort to reduce the company's own emissions profile. Along with Russia's Rosneft, Lukoil is also a lead explorer in Kazakhstan's Karachaganak and Tengiz oil and gas fields.

The Kazakh authorities are also in talks with Russia's state-owned firm Rosatom for the possible construction of a nuclear power plant. Russia is thus using its role as the leading exporter of civilian nuclear reactors to boost its geopolitical influence and shape the rules in the sector (Bordoff 2020). The government of Uzbekistan is also working with Rosatom to add nuclear to its energy balance along with RE.

Turkey is also emerging as a key energy partner for the region. Ankara has long aspired to operate as a gateway and power broker of Eurasian energy supplies. Uzbekistan has recently announced several investments in combined cycle gas turbine (CCGT) power generation assets in the south of the country, with crucial funding from Turkey. And in September 2020, it was announced that Turkey's Cengiz Energy will establish a natural gas combined cycle power plant in Uzbekistan with an investment of US$150 million (AA Energy 2020).

6.1 Intra-Regional Cooperation

Given that the survival of unpopular oil and gas-reliant elites still depends on the sale of hydrocarbons, we have recently witnessed a rekindled interest *among* Central

Asian hydrocarbon producers too, in forging closer bonds for joint oil and gas investment and trade.

In a remarkable sign of tightening relations, Iran, Azerbaijan and Turkmenistan signed a trilateral natural gas swap deal on the sidelines of the 15th summit of the Economic Cooperation Organization (ECO), in Ashgabat, on 27 November 2021 (Eurasia Review 2022). Iran and Turkmenistan had previously been locked in a gas dispute that hindered cooperation. Azerbaijan and Turkmenistan, too, have resolved their competing claims over oil and gas fields located in the Caspian Sea. Iran cites the pursuit of more active 'gas diplomacy' as the reason for a reconciled relationship with Turkmenistan.

At the same time, part of the driving force behind resolving their long-lasting gas dispute comes from Turkmenistan. As the sixth-largest holder of gas reserves in the world, Ashgabat is pursuing a gas export route diversification policy in four geographical directions: to the North (Russia), East (China), South (Afghanistan and Pakistan), and West (Iran, the Caspian Sea, and Azerbaijan). Turkmenistan, however, faces significant challenges in implementing this strategy. Europe's ambitious 'greening' policies mean that a Trans-Caspian pipeline to supply Turkmen gas to Europe is unlikely, while the country's diversification of revenue sources has stalled. In such circumstances, the resumption of the Turkmenistan-Iran gas agreement will allow Ashgabat to raise hydrocarbon revenues as well as diminish reliance on China, its main gas export market. In line with this diversification drive, President Berdymuhamedow has recently expressed readiness to renew partnerships with Russia in the sphere of oil and gas in the Caspian Sea (Caspian News 2021b).

In sum, given the central role that the oil and gas sectors continue to play in the region's economies, the objective of maximizing fossil fuel revenues will continue to rank highly in any policy decision and act as a constraint on foreign policy choices. Essentially, the oil and gas-producing nations of Central Asia will be more prone to foreign policy volatility to compensate for declining hydrocarbon revenues.

7 Conclusions

Given their heavy reliance on fossil fuel rents, Central Asian petrostates might find it difficult to pursue policies that jeopardize the sectional interests of economic elites. As a result, energy transition is a delicate tightrope for the leaders of these three petrostates to walk. They understand that the elites to which they are beholden rely on fossil fuels, but simultaneously recognize that the future of their economies—and the global economy—depends on renewable energy. Scholarly analysis has not, as yet, systematically addressed the interactions between powerful informal elites and government agendas in pushing through socio-economic reforms in critical transition periods.

This chapter has therefore sought to provide a more nuanced understanding of the interactions between informal elites, the state and new transnational partners in three crucial petrostates as we enter a post-oil era in world politics. In this new era, the hydrocarbon producers of Central Asia are in a less favourable position. This research, however, suggests that, at least compared to the early years of independence, Central Asian petrostates have both an increased motivation and a greater capacity to 'go abroad' in their search for fossil fuel revenues to balance their budgets.

One of the most interesting—and under-studied—effects of the global energy transition is precisely this tangled web of vested interests and addiction to fossil-fuel sales that have bound [Central Asian] hydrocarbon producers ever more tightly to their primary export. Thus, instead of reducing economic overdependence on hydrocarbons—a seemingly logical step, given the inexorable decline in global fossil fuel demand—the energy transition has directly affected foreign policy choices and spurred a desperate search for new hydrocarbon customers.

Thus far, the energy transition has created more volatile and unpredictable foreign policy profiles across the region driven mainly by opportunistic relationships that will last for generations to come. Future research should look deeper into complexities, such as the institutional relationships between the state and elite actors, and how these relationships can be implicated both in domestic political struggles and in the search for new transnational opportunities that the energy transition is unleashing.

References

AA Energy (2020) Cengiz Energy invests $150M in Uzbekistan power plant, September 21
Acemoglu D, Robinson JA (2006) Economic origins of dictatorship and democracy. Cambridge University Press, Cambridge
Beblawi H, Luciani G (eds) (1987). The rentier state: Essays in the political economy of Arab countries. Croom Helm, New York
Blondeel M, Colgan J, Van de Graaf T (2019) What drives norm success? Evidence from anti-fossil fuel campaigns, Glob Environ Polit 19:4, November
Bloomberg (2021) Shell loses climate case that may set precedent for big oil, May 26
Bordoff J (2020) Everything you think about the geopolitics of climate change is wrong. Foreign Policy, October 5
BP Statistical Review of World Energy (2021) Primary energy
Caspian News (2021a) Russia, Kazakhstan push for hydrocarbon expansion with projects in Caspian, October 2
Caspian News (2021b) Turkmen leader, Russia's Lukoil head discuss energy projects, October 23
Dudley D (2020) Gulf power developers take an early lead in Uzbekistan's renewable energy drive. Forbes, March 23
Energy Central News (2021) Kazakhstan: 3% of all energy generated by renewable sources in 2020, February 10
Eshchanov B et al (2019a) Renewable energy policies of the Central Asian countries. Cent Asia Reg Data Rev 16:1–4
Eshchanov B et al (2019b) Wind power potential of the Central Asian countries. Cent Asia Reg Data Rev 17:1–7
Eshchanov B et al (2019c) Solar power potential of the Central Asian countries. Cent Asia Reg Data Rev 18:1–7
Eshchanov B et al (2019d) Hydropower potential of the Central Asian countries. Cent Asia Reg Data Rev 19:1–7
Eurasianet (2020) Chinese solar investments in Central Asia: a snapshot, October 15

Eurasianet (2021) Central Asia courts green energy investors, April 15
Eurasia Review (2022) New gas cooperation between Iran and Turkmenistan: prospects and limitations, January 4
Fattouh B (2021) Saudi oil policy: continuity and change in the era of the energy transition. OIES Paper: WPM 81, January
Fattouh B, Poudineh R, West R (2019) The rise of renewables and energy transition: what adaptation strategy exists for oil companies and oil-exporting countries? Energy Transit 3:45–58
Laldjebaev M, Isaev R, Saukhimov A (2021) Renewable energy in Central Asia: an overview of potentials, deployment, outlook, and barriers. Energy Reports 7:3125–3136
Koch N, Tynkkynen V-P (2019) The geopolitics of renewables in Kazakhstan and Russia. Geopolitics 26:1–20
Kudaibergenova DT, Laruelle M (2022) Making sense of the January 2022 protests in Kazakhstan: failing legitimacy, culture of protests, and elite readjustments. Post-Soviet Affairs
Morrison K (2009) Oil, nontax revenue, and the redistributional foundations of regime stability. Int Organ 63:107–138, Winter
O'Sullivan M, Overland I, Sandalow D (2017) The geopolitics of renewable energy. Belfer Center Working Paper, Cambridge, MA
Reuters (2019) Kazakhstan to divert some oil flows from Europe to China, July 3
Reuters (2021) Big oil may get more climate lawsuits after Shell ruling, May 28
Scholten D, Bazilian M, Overland I, Westphal K (2020) The geopolitics of renewables: new board, new game. Energy Policy 138:111059
Shadrina E (2020) A double paradox of plenty: renewable energy deployment in Central Asia. Eurasian Geogr Econ 63(1):1–26
Shehabi M (2019, April) Diversification in Gulf hydrocarbon economies and interactions with energy subsidy reform: lessons from Kuwait, OIES PAPER: MEP 23, OIES: The Oxford Institute for Energy Studies
Skalamera M (2017) Russia's lasting influence in Central Asia. Survival 59(6):123–142
Skalamera M (2020) The 2020 oil price dive in a carbon-constrained era: strategies for energy exporters in central Asia. Int Aff 96(6):1623–1642
Skalamera Groce M (2020) Circling the barrels: Kazakhstan's regime stability in the wake of the 2014 oil bust. Cent Asian Surv 39:480–499
Vakulchuk R, Overland I (2021) Central Asia is a missing link in analyses of critical materials for the global clean energy transition. One Earth 4(12):1678–1692
Van de Graaf T, Bradshaw M (2018) Stranded wealth: rethinking the politics. Int Aff 94(6):1309–1328
Vestnik Kavkaza (2022) Kazakhstan turning energy hub for China, Europe, January 1
Wishart A, Abidi A (2021) The energy transition in Central Asia: drivers, policy and opportunities. Constr Law Int 16:4, December

Morena Skalamera is Assistant Professor at Leiden University and an associate at the Belfer Center, Harvard Kennedy School. She teaches courses in international political economy, with a regional focus on Russia and Eurasia. Her research interests include the political economy of Eurasia, Russian and post-Soviet Politics, and the Geopolitics of Energy in Eurasia. Dr. Skalamera has spent extensive time conducting field research in post-Soviet Eurasia, especially in Russia and Central Asia, and in Turkey. Her current writings focus on issues of identity politics, exploring the particular interplay between international and domestic factors in policy making, and contemporary state–market relations with a focus on energy policy dynamics in Russia and Central Asia. She is currently working on a book manuscript that examines how energy firms have shaped the energy relationship between Russia and Europe and the energy strategies of the former Soviet states.

Open Access This chapter is licensed under the terms of the Creative Commons Attribution 4.0 International License (http://creativecommons.org/licenses/by/4.0/), which permits use, sharing, adaptation, distribution and reproduction in any medium or format, as long as you give appropriate credit to the original author(s) and the source, provide a link to the Creative Commons license and indicate if changes were made.

The images or other third party material in this chapter are included in the chapter's Creative Commons license, unless indicated otherwise in a credit line to the material. If material is not included in the chapter's Creative Commons license and your intended use is not permitted by statutory regulation or exceeds the permitted use, you will need to obtain permission directly from the copyright holder.

Towards a Geoeconomics of Energy Transition in Central Asia's Hydrocarbon-Producing Countries

Yana Zabanova

Abstract The global energy transition and the growing ambition of decarbonisation policies in the world's leading economies are bound to affect Central Asia's three hydrocarbon producers, yet there has been little research on the issue. This chapter uses the theoretical toolbox of geoeconomics to analyse the implications of energy transition for Kazakhstan, Uzbekistan and Turkmenistan across four dimensions: resources, energy infrastructure, strategic industries and clean energy technologies and the rules of international economic interaction. Kazakhstan remains the region's renewable energy frontrunner, while Turkmenistan is yet to make its first steps. Despite a late start, Uzbekistan is benefiting from the falling cost of renewable energy technologies and has attracted major industry players. Rather than pursuing fossil-fuel phaseout, these countries seek to capture more value from their hydrocarbon resources—especially natural gas—while adding renewables to the mix to bolster energy security and help decarbonise the economy. However, infrastructure constraints remain an obstacle to low-carbon energy production. Lacking the influence to shape the international rules of engagement on energy transition, the three countries are employing a variety of strategies to manage their asymmetrical dependence on key partners—be they China, the EU or Russia—and to diversify their international engagement.

Keywords Geoeconomics · Energy transition · Central Asia

1 Introduction

Traditionally, the geopolitics of energy has focused on oil and gas as a source of power and influence in the global economy. More recently, however, increasing attention has been paid to energy transition and its geopolitical ramifications. These include the

Y. Zabanova (✉)
Research Institute for Sustainability (RIFS)—Helmholtz Centre, Potsdam, Germany
e-mail: yana.zabanova@rifs-potsdam.de

declining power of hydrocarbon exporters, an expected reduction in the potential for conflict in a renewables-dominated world, and the intensifying competition between China and the West on clean energy technologies. Furthermore, there are growing security risks associated with critical minerals needed for the energy transition and in the cybersecurity domain but also new economic opportunities for countries of the Global South that are rich in renewable energy and critical raw materials (Vakulchuk et al. 2020).

In parallel, there has been a growing interest in *geoeconomics* as a related but different lens through which to observe power relations within the global economy. In his seminal 1990 article 'From Geopolitics to Geo-Economics', Edward Luttwak argued that, with the collapse of the bipolar world order, nation-states were increasingly relying on economic instruments rather than military might to achieve their geostrategic goals. The emergence of global value chains and the rise of China as the world's economic powerhouse have further increased the salience of geoeconomics. The security implications of supply chain disruptions became painfully clear during the Covid-19 pandemic and in the wake of the Russian invasion of Ukraine in February 2022.

Central Asia has rarely been mentioned in these analyses; this reflects the near total absence of academic research on climate change and decarbonisation in Central Asia more widely (Vakulchuk et al. 2022). Koch and Tynkkynen (2021) use a critical geopolitics approach to compare renewable energy development in Kazakhstan and Russia, while Vakulchuk and Overland (2021) analyse Central Asia's potential to supply critical materials for clean energy technologies. To the author's knowledge, no contributions have addressed the geoeconomics of the energy transition in Central Asia. This is the research gap that this chapter is aiming to address.

In the global energy transition, Central Asia remains a laggard. Wind and solar power deployment in the region began only around 2013 and for a long time took place almost exclusively in Kazakhstan. As of mid-2022, the total installed capacity was still below 2 GW. There is much room for growth: the technical solar power potential of Central Asian countries exceeds their current power generation levels by a factor of twenty (Eshchanov et al. 2019b). For wind power, the potential is even higher, with 70% of this concentrated in Kazakhstan (Eshchanov et al. 2019a). Yet, there are many challenges ahead. Mass deployment and integration of renewable energy require a large-scale upgrade of the region's ailing grid infrastructure, much of it dating back to Soviet times. Energy tariffs are low, resulting in high levels of energy intensity and a lack of investment in electricity grids, as well as distorted price signals, making it difficult for renewables to compete with traditional—heavily subsidised—power generation.

2 Conceptual Framework: Towards a Geoeconomics of Energy Transition

There is no uniform definition of geoeconomics, and the term is often confusingly used alongside geopolitics. Wigell (2016, 135) defines geoeconomics as 'the geostrategic use of political power by economic means', while Blackwill and Harris (2016, 9) describe it as 'the use of economic instruments to promote and defend national interests, and to produce beneficial geopolitical results; and the effects of other nations' economic actions on a country's geopolitical goals'. Diesen (2019, 568) employs the concept of 'asymmetrical dependence', which 'enables the more powerful and less dependent state to set favourable conditions for economic cooperation and to extract political concessions from the more dependent state'. Geoeconomics combines an outward pursuit of influence in the global economy or in a given region with a more inward-oriented approach to strengthening one's economic resilience. The latter may be particularly relevant for states that lack the capacity to actively project power internationally.

With its nuanced treatment of the economic interdependencies created by global value chains—where some countries are more indispensable than others—and the resulting power relations, geoeconomics is a promising lens through which to analyse the global energy transition. Renewable-rich regions may become attractive locations for industries seeking to decarbonise; imports of hydrocarbons may be superseded by imports of renewable energy, critical materials or green intermediary industrial products, and lead to the emergence of new strategic partnerships; and governments and private actors may compete—or cooperate—in creating new value chains for climate-friendly technologies and products. Drawing on Diesen (2019) and Wigell (2016), this chapter explores the following dimensions of the geoeconomics of energy transition: *resources, energy infrastructure, strategic industries and clean energy technologies, and the rules of international economic interaction.*

Despite the more even distribution of renewable energy sources as compared to oil and gas, *resources* will retain importance in a net zero world: sites with high RE endowments will be needed to decarbonise industrial production; manufacture of clean energy technologies will require ever growing volumes of critical minerals—whose supply is even more concentrated geographically than that of hydrocarbons; water availability will be crucial for renewable hydrogen production through electrolysis; CO_2 geological storage sites will become an important asset; and finally, natural gas is likely to preserve its strategic importance for a long time to come.

Energy infrastructure will remain essential as well. This includes pipelines which will be used to transport clean hydrogen or hydrogen-methane blends, seaports ready for handling sustainable liquid fuels and hydrogen, and tube trailers for road transportation. With the electrification of new sectors—such as transport, industrial processes and residential heating and cooling (and the corresponding increase in the disruptive potential of blackouts)—and with the growing share of intermittent renewables, there will be a need for a better quality grid infrastructure and more regional interconnections. Furthermore, grid infrastructure can be used 'as a means of projecting political power and authority beyond territorial space' (Grinschgl et al. 2021, 7).

Strategic industries and clean energy technologies constitute a prime area of geoeconomic competition, especially among powerful actors like China, the US and the EU, who fund R&D programmes and promote clean technology manufacturing. Less influential states may try to position themselves along certain segments of clean energy value chains or extract benefits for domestic industrial development from technology and R&D cooperation with leading countries. In addition, petrostates may seek to capture more added value from oil and gas by specialising in high-carbon applications, e.g. by developing the petrochemical industry or increasing oil refining capacity (Goldthau and Westphal 2019).

Finally, ***the rules of international economic interaction*** include, among other things, a wide array of climate policies, measures to prevent carbon leakage, green investment taxonomies governing access to climate finance, sustainable product standards, certification procedures and life cycle assessment methodologies for calculating the CO_2 footprint. Such rules are predominantly negotiated by powerful actors, while others are left with finding ways to adapt. Adopting the 'green' rules of the game may also become a competitive advantage in the global markets, as well as a reputational dividend.

The following sections explore the geoeconomics of the energy transition in Kazakhstan, Uzbekistan and Turkmenistan, focusing on the four dimensions described above.

3 Kazakhstan

Kazakhstan (population 18.7 million), Central Asia's wealthiest nation, is the epitome of a hydrocarbon-dependent economy. Its oil exports are the main source of revenue for the state budget, it produces and exports natural gas, and it relies on cheap domestic coal for power and heat generation. Yet, Kazakhstan's leadership was early to recognise the importance of clean technologies for economic competitiveness and international prestige. Kazakhstan became the region's first country to actively develop its renewable energy sector, launching an emissions trading system in 2013 and announcing a carbon neutrality target (for 2060). In 2013, Kazakhstan adopted its Concept of Transition to a Green Economy, which contains renewable energy targets (3% by 2020, 10% by 2030—increased to 15% in 2022 and 50% by 2050) and introduced a feed-in tariff for renewable energy (replaced by renewable energy auctions in 2018). By early 2022, Kazakhstan had installed 684 MW of wind power, 1 GW of solar, 281 MW of small hydropower and 8 MW of biogas (Ministry of Energy 2022). However, RE development in Kazakhstan has decelerated, owing to grid integration difficulties.

3.1 Resources

In the age of energy transition, the strategic significance of hydrocarbon resources is changing. The expected fall in the global demand for oil will decrease Kazakhstan's export revenues. Coal, despite its abundance in Kazakhstan, is no longer expected to play a significant role in new power generation capacity owing to societal opposition related to environmental issues and a lack of access to international finance. Natural gas, however, is viewed positively as the most feasible option to replace old coal-fired facilities, as well as to provide a much-needed balancing capacity for the Kazakh power system.

Kazakhstan's significant RE endowment can become a strategic resource to help decarbonise its energy-intensive, export-oriented economy; the mining and metals industry is a prominent example. The country has also begun looking for ways to situate itself in the emerging international hydrogen economy. The national investment agency Kazakh Invest has signed a memorandum of understanding with Svevind Energy, a German company, to install up to 30 GW of electrolyser capacity to produce renewable hydrogen and ammonia in Mangistau Region in western Kazakhstan. In addition, Kazakhstan's state-owned oil and gas company KazMunayGas has signed an agreement with Linde, a leading German industrial gas company, to develop hydrogen projects.

Kazakhstan is also rich in critical raw materials. Some big players have turned to Kazakhstan to explore its rare earth production potential as a way to reduce their own asymmetrical dependence on China. In 2012, Japan's Sumitomo Corporation launched a joint venture with Kazakhstan's state-owned nuclear company Kazatomprom to produce heavy rare earth metals at a plant in Stepnogorsk, northern Kazakhstan. The cooperation soon petered out, but there is now, however, a renewed interest in Kazakhstan in developing critical minerals, including lithium. In addition, Kazakhstan is the world's largest producer and exporter of natural (non-enriched) uranium, with a share of over 50% of the global market. It may potentially gain importance if the West decides to sanction Russian uranium.

3.2 Energy Infrastructure

Currently, Kazakhstan is considering ways in which its gas transportation infrastructure can be used to transport clean hydrogen or ammonia in the future (the original plans to use the Russian gas transportation system have been put in question by the war in Ukraine). As for Kazakhstan's power system, it suffers from the poor condition of the grid, weak connections between the north and the south, and a completely isolated western zone. This is already becoming a serious liability in meeting domestic demand and integrating the growing amount of power generated from renewables. In recent years, Kazakhstan has increasingly relied on relatively expensive electricity imports from Russia to balance its system and satisfy growing domestic demand, viewing this condition as a structural vulnerability.

3.3 Strategic Industries and Clean Energy Technologies

Kazakhstan has started producing nuclear fuel assemblies at a plant in Ust-Kamenogorsk launched in November 2021 as a result of a joint venture between Kazatomprom and the state-owned China General Nuclear Power Group. The plant uses French technology and Russian-enriched uranium and plans to export fuel assemblies to China. In parallel, there has been a prolonged discussion in the country on whether to build a nuclear power plant (a project that Russia's state nuclear energy corporation Rosatom has been eyeing, although Kazakhstan remains open to other technology suppliers as well), despite strong societal opposition. Kazakhstan's 2060 carbon neutrality target has been used as a justification for these plans. On the one hand, a nuclear power plant could allow Kazakhstan to capture added value from its uranium reserves and improve energy security. On the other, it would also make Kazakhstan dependent on external suppliers of technology and skilled labour.

Other efforts to move up the value chain in clean energy technologies have had limited success. As part of its State Industrialisation Programme (2010), Kazakhstan has attempted to develop a fully vertically integrated cycle of PV module manufacturing using domestic silicon. A solar PV module manufacturing plant, Astana Solar, was launched in 2012, yet, despite state support, soon found itself unable to compete with cheap Chinese PV modules, eventually ceasing production. Today, nearly all solar PV modules used in Kazakhstan are imported from China.

3.4 Rules of International Economic Interaction

In the geoeconomics of energy transition, the rules of international interaction are largely developed by the leading economic powers, with other actors having little say in shaping these. The EU's proposed Carbon Border Adjustment Mechanism (CBAM) to prevent carbon leakage is a case in point. Kazakhstan, for whom the EU is a key trading partner, will be affected by CBAM rules. This has led to calls within Kazakhstan to bring its currently dysfunctional emissions trading system in line with EU rules.

Another relevant example relates to local content requirements (LCRs). Unlike Russia, Kazakhstan has never introduced LCRs, which would have obligated investors to use at least some Kazakhstani-made components, in its renewable energy tenders. This may be a reflection of the influence of the European Bank of Reconstruction and Development (EBRD), which has been one of the key players in RE development in Kazakhstan, both acting as a financing institution and consultant to the government on RE regulations. The EBRD has been strongly against LCRs, arguing that they would complicate project completion and result in higher costs. Membership of the World Trade Organisation (WTO), which Kazakhstan joined in November 2015, likely played a role too given that WTO members are required to treat domestic and imported products equally.

4 Uzbekistan

Uzbekistan, Central Asia's most populous country (population 35 million) and an important gas producer, is intent on challenging Kazakhstan's 'green' leadership in the region. Uzbekistan was late in adopting a law on renewable energy: it did so only in 2019, a decade after Kazakhstan. Similarly, Uzbekistan announced its net zero target later than Kazakhstan—in January 2021—but it also chose a more ambitious one, aiming for 2050 rather than 2060. The relatively late start has allowed Uzbekistan to benefit from the falling costs of RE technologies. This has translated into renewable energy auction prices that are lower than any that have been achieved in Kazakhstan. In Uzbekistan's first competitive solar tender in October 2019, the winner was the United Arab Emirates' Masdar, offering 0.027 USD per kWh—a record low for the region at the time. The tender was for the 100 MW Nur-Navoi plant in the north of the country—Uzbekistan's first utility-scale solar power plant—which was inaugurated in August 2021.

Uzbekistan's interest in renewable energy is to a large extent driven by energy security considerations. Although nominally self-sufficient in energy, Uzbekistan is a country with a young and rapidly growing population. There is a clear need for increased domestic energy production to match the rising demand, and renewables are an important piece of the puzzle.

4.1 Resources

Uzbekistan currently uses most of its gas production to supply domestic demand, both for power and heat generation and for producing gas motor fuel. Yet Uzbekistan's proven gas reserves are limited and may not last much longer than another 25–30 years; there is therefore a desire to monetise these in the most profitable way. For this reason, Uzbekistan is planning to boost its gas production, mainly for export to China, and is already developing gas processing industries to add value to its gas production, with further plans to develop its domestic petrochemical industry.

Unlike Uzbekistan's dwindling gas reserves, its solar potential is enormous and remains largely untapped. In May 2019, Uzbekistan adopted a *Renewable Energy Law* granting subsidies and incentives to RE facilities and RE manufacturers. In May 2020, the *Concept of Providing the Republic of Uzbekistan with Electric Energy for 2020–2030* set out ambitious plans for the overhaul of the power sector. These include more than doubling the total installed capacity from 12.9 GW to 29.3 GW by 2030 and increasing the share of renewables (including hydro) in power generation to at least 20% by 2025 and at least 25% by 2030. This is to be achieved through the construction of 5 GW of solar, 3 GW of wind and 2 GW of hydropower capacity. In 2021, the Uzbek Ministry of Energy announced plans to raise these targets to 7 GW solar and 5 GW wind (Ministry of Energy 2021). Uzbekistan has prioritised the construction of large-scale RE plants (up to 500 MW) in the form of public–private partnerships with renewable energy market leaders such as Total Eren (France), Masdar (UAE) and Saudi Arabia's ACWA Power, receiving technical assistance from the International Finance Corporation (IFC).

The *Concept* also foresees the construction of Central Asia's first nuclear power plant with a capacity of 2.4 GW, to be built by Rosatom. Like Kazakhstan, Uzbekistan could use its own uranium reserves to produce fuel. However, the future of this project has been rendered uncertain in the wake of Russia's invasion of Ukraine and Russia's resulting geopolitical isolation. Finally, Uzbekistan is also investing in R&D in hydrogen technologies. It has set up a National Research Institute for Renewable Energy, which includes a Centre for Hydrogen Research and is seeking to develop technology cooperation on hydrogen with China, Russia and the EU. As of 2022, Uzbekistan's government is also working on a hydrogen strategy in close cooperation with industry leaders that are already investing in the Uzbek energy sector: ACWA Power, Siemens and the US industrial gas company Air Products. In addition, in 2021, Uzbekistan requested that the World Bank explore potential options for producing blue hydrogen (from natural gas) and developing carbon capture and storage (CCS) technology solutions in Uzbekistan.

4.2 Energy Infrastructure

Uzbekistan has been exporting a smaller part of its gas products, mainly to China, Russia and Kazakhstan. It is now considering retrofitting its gas pipelines to transport clean hydrogen in the future. As a doubly landlocked nation, Uzbekistan has no seaports and would be unable to participate in future hydrogen or ammonia trade by ship, which would expand the range of potential buyers.

As for the Uzbek power system, it is in poor condition, suffering from high losses and lacking the capacity to meet the demand of a rapidly growing population. Frequent electricity and gas shortages are the cause of much public discontent. At the moment, Uzbekistan is chiefly focused on modernising its own domestic energy infrastructure, which is a precondition for reducing losses and integrating growing amounts of renewable energy.

4.3 Strategic Industries and Clean Energy Technologies

The government of Uzbekistan is clearly interested in capturing more added value along the energy value chain. For instance, Uzbekistan is planning to build on its strengths in the automotive industry and the experience of using compressed natural gas for a large share of its light vehicle fleet to develop clean hydrogen transportation. Another priority area is the chemical and petrochemical industry, where the government already has significant processing projects underway and is planning further investment—over 12 billion USD between 2019 and 2030—which would allow it to capture more value from its diminishing gas reserves. As one of the world's most energy-intensive economies, Uzbekistan also has a long way to go in decarbonising its industrial sector. Uzbekistan's renewable energy resources—and possibly clean hydrogen—could play a role in this effort.

4.4 Rules of International Economic Interaction

Like Kazakhstan, Uzbekistan may not be able to influence the rules of international economic interaction in the area of the energy transition, but it is trying to position itself as Central Asia's 'green champion' to attract foreign investment and technical expertise. It is also seeking to extract benefits from the geopolitical and geoeconomic strategies of more powerful countries and is pursuing multi-vector cooperation with key actors. Interested in large-scale renewable energy deployment at a low cost, Uzbekistan has opted for close cooperation with Masdar, an energy company controlled by the UAE's Mubadala Investment Company, which can be characterised as a sovereign wealth fund. The latter has strategic goals to increase its presence in Eurasia and is less profit-oriented than regular private investors, thus being able to offer especially low tariffs. Likewise, when it comes to financing clean energy, Uzbekistan has been able to benefit from large amounts of funding supplied by international development institutions such as the Asian Development Bank, the EBRD and the IFC. In 2019, Uzbekistan became the first non-African country to secure a place in the IFC's Scaling Solar programme, which helps governments to develop large-scale solar projects with the participation of private lenders. Finance from international financial institutions does, however, come with its own conditionalities, i.e. economic, social or environmental rules, which need to be followed.

In its game of catch-up, Uzbekistan is relying on international expertise in developing nearly all its sustainability-focused programmes. For instance, Uzbekistan's Ministry of Energy and Ministry of Investment and Foreign Trade have engaged such leading consultancies as Corporate Solutions, Tractebel and Guidehouse, with support from the government of Japan and the EBRD, to develop a roadmap for a carbon–neutral electricity sector by 2050. A similar process is currently being pursued to draft the national hydrogen strategy. The EBRD is assisting Uzbekistan, which joined the Global Methane Pledge in 2022, with developing a national methane emissions programme to bring these emissions down by 30% by 2030.

5 Turkmenistan

Turkmenistan (population 6 million), Central Asia's reclusive and authoritarian gas giant, has so far been an idle onlooker in the global energy transformation. The power transition from President Gurbanguly Berdymukhamedov to his son Serdar in 2022 raised hopes that Turkmenistan will open to the international community and increase its cooperation on climate policy and energy transition. As of 2022, no renewable energy projects have been implemented in the country; however, there are some cautious signs that this might soon change.

5.1 Resources

Turkmenistan may possess some of the world's largest gas reserves—fourth after Russia, Qatar and Iran—as well as some oil, but this hydrocarbon wealth has not materialised prosperity for the Turkmen population. It has, however, turned Turkmenistan into possibly the most gas-dependent country in the world when it comes to both its energy sector and its export structure, with an utterly undiversified, non-resilient and CO_2-intensive economy.

With its wind-swept Caspian shores, vast deserts and excellent solar radiation levels, Turkmenistan also has a significant renewable energy potential and no shortage of available land. This has not yet translated into active renewable energy deployment, but there are indicators of growing interest. In December 2020, Turkmenistan adopted the *National Strategy for Renewable Energy Development* until 2030, followed by the *Renewable Energy Law* in March 2021, both drafted with support from the UN Development Programme (UNDP); more regulations are currently in the pipeline. In October 2021, Turkmenistan signed a Memorandum of Understanding on renewable energy with Masdar, which is actively expanding its presence in post-Soviet Eurasia and is known for delivering utility-scale RE projects at a low price. The Asian Development Bank is supporting an innovative project in Turkmenistan where concentrated solar power technologies would be installed at existing gas-fired power plants whose turbines could then run not only on gas but also on solar power whenever available (Asian Development Bank 2021).

5.2 Energy Infrastructure

Turkmenistan is very limited in the range of available gas export routes and can sell gas to very few customers. Until 2008, its main importer was Russia, but since 2009, following the construction of a new gas pipeline, it has been China, giving the latter significant leverage over prices and placing Turkmenistan in the position of strong asymmetrical dependence. Turkmenistan, however, also has two Caspian Sea LNG terminals which have allowed it to expand its export geography to as far away as Japan. Given that future hydrogen market development may share a great deal of similarities with LNG, technical competencies in this area may help Turkmenistan position itself as a blue or green hydrogen exporter.

5.3 Strategic Industries and Clean Energy Technologies

Like Uzbekistan, Turkmenistan has made some effort to invest in gas processing so as to better monetise its gas reserves and move up the value chain. A 3.4 billion USD gas chemical complex opened in Kiyanly, western Turkmenistan, in 2018. In 2019, Turkmenistan launched the world's first gas-to-gasoline plant in the Ahal Region. The plant uses Danish Haldor Topsoe's innovative technology which enables the production of especially pure gasoline. In 2021, Haldor Topsoe signed a memorandum of understanding with Turkmenistan to construct a gas chemical plant to produce ammonia and methanol. If these fuels are decarbonised, Turkmenistan has a chance to become a sustainable fuel producer.

By and large, however, Turkmenistan has not yet taken steps towards adapting its carbon-dependent economy to the decarbonising world. More active participation in the energy transformation would give Turkmenistan a lifeline by diversifying its economy (and thereby increasing its resilience, which is a geoeconomic imperative). Technology cooperation with industry leaders—such as Masdar or Haldar Topsoe—is a step in the right direction. In addition, there is hope that China's own green transition plans, coupled with its outsized influence in Turkmenistan, will have spillover effects in this Central Asian country; after all, China is known for successfully completing infrastructure projects in the most politically risky locations.

5.4 Rules of International Economic Interaction

Turkmenistan prides itself on 'neutrality', a principle embedded in its Constitution. In practical terms, this, together with its repressive political regime, has translated into high levels of self-imposed isolation, including a striking lack of involvement in the global climate agenda. However, cooperation on energy transition and sustainability might be one of the few promising ways for Turkmenistan to improve its international image and diversify its economy. This was likely the country's motivation for becoming a member of IRENA in 2018, cooperating with the UNDP on renewable energy legislation, and participating in various sustainability-oriented initiatives.

6 Conclusion

Central Asia's hydrocarbon producers are increasingly grappling with ways to adapt to the accelerating global energy transformation. While not influential enough to actively project geoeconomic power or to shape the global climate agenda, these countries are interested in managing their asymmetrical dependences and raising the resilience of their carbon-intensive economies. Kazakhstan, Uzbekistan and Turkmenistan seek to monetise their hydrocarbon reserves by moving up the value chain and branching into the chemical industry, or by finding new export destinations. In addition, they are beginning to view their renewable energy potential as a strategic resource that can help attract foreign investment, bolster energy security, help decarbonise energy-intensive exports, and improve their international image. Uzbekistan and Kazakhstan are investigating the possibility of producing low-carbon and renewable hydrogen or ammonia for domestic use and for export. There is also renewed interest in expanding the production of materials critical for the energy transition. Finally, all three countries are trying to draw benefits from the geoeconomic and geopolitical strategies of other key players, such as China, the EU, the UAE and Russia, using them as sources of foreign investment, clean energy finance, technological cooperation and technical and regulatory expertise.

References

Asian Development Bank (2021) Game-changer: can solar energy be generated in old gas power plants?, August 11. https://blogs.adb.org/blog/game-changer-can-solar-energy-be-generated-old-gas-power-plants

Blackwill R, Harris J (2016) War by other means. Harvard University Press, Cambridge, MA

Diesen G (2019) The geoeconomics of Russia's greater Eurasia initiative. Asian Polit Policy 11(4):566–585. https://doi.org/10.1111/aspp.12497

Eshchanov B, Abylkasymova A, Aminjonov F, Moldokanov D, Overland I, Vakulchuk R (2019a) Wind power potential of the central asian countries. Cent Asia Reg Data Rev 17(2019):1–7

Eshchanov B, Abylkasymova A, Aminjonov F, Moldokanov D, Overland I, Vakulchuk R (2019b) Solar power potential of the Central Asian countries. Cent Asia Reg Data Rev 18(2019):1–7

Goldthau A, Westphal K (2019) Why the global energy transition does not mean the end of the petrostate. Global Policy

Grinschgl J, Pepe J, Westphal K (2021) A new hydrogen world: geotechnological, economic, and political implications for Europe. SWP Comment 2021/C 58, December 16. https://doi.org/10.18449/2021C58

Koch N, Tynkkynen V (2021) The geopolitics of renewables in Kazakhstan and Russia. Geopolitics 26(2):521–540. https://doi.org/10.1080/14650045.2019.1583214

Luttwak EN (1990) From geopolitics to geo-economics: logic of conflict, grammar of commerce. The National Interest, No. 20

Ministry of Energy of the Republic of Kazakhstan (2022) Razvitiye vozobnovlyayemykh istochnikov energii (Development of renewable energy sources). https://www.gov.kz/memleket/entities/energo/activities/4910?lang=ru

Ministry of Energy of the Republic of Uzbekistan (2021) Uzbekistan's Ministry of Energy plans to increase its 2030 renewables targets. https://minenergy.uz/en/news/view/1389

Vakulchuk R, Overland I (2021) Central Asia is a missing link in analyses of critical materials for the global clean energy transition. One Earth 4:1678–1692. Available at SSRN: https://ssrn.com/abstract=3989008

Vakulchuk R, Overland I, Scholten D (2020) Renewable energy and geopolitics: a review. Renew Sustain Energy Rev 122(C):109547

Vakulchuk R, Daloz A, Kverland H, Sagbakken F, Standal K (2022) A void in Central Asia research: climate change. Cent Asian Surv. https://doi.org/10.1080/02634937.2022.2059447

Wigell M (2016) Conceptualizing regional powers' geoeconomic strategies: neo-imperialism, neo-mercantilism, hegemony, and liberal institutionalism. Asia Eur J 14:135–151

Yana Zabanova is a research associate at the Research Institute for Sustainability (RIFS) in Potsdam, Germany, where she focuses on energy transitions in Central Asia and Russia, the emerging international hydrogen economy and, more generally, on the geopolitics and geoeconomics of the global energy transformation. Yana is also a PhD candidate at the University of Groningen (Netherlands). As part of her doctoral research, she studies renewable energy and clean hydrogen development in Eurasian hydrocarbon-rich countries such as Russia and Kazakhstan from a comparative perspective.

Open Access This chapter is licensed under the terms of the Creative Commons Attribution 4.0 International License (http://creativecommons.org/licenses/by/4.0/), which permits use, sharing, adaptation, distribution and reproduction in any medium or format, as long as you give appropriate credit to the original author(s) and the source, provide a link to the Creative Commons license and indicate if changes were made.

The images or other third party material in this chapter are included in the chapter's Creative Commons license, unless indicated otherwise in a credit line to the material. If material is not included in the chapter's Creative Commons license and your intended use is not permitted by statutory regulation or exceeds the permitted use, you will need to obtain permission directly from the copyright holder.

Local Climate Change Impacts
and Adaptation in Central Asia

The Dual Relationship Between Human Mobility and Climate Change in Central Asia: Tackling the Vulnerability of Mobility Infrastructure and Transport-Related Environmental Issues

Suzy Blondin

Abstract Human mobility impacts the global climate and the climate in turn impacts human mobility. Fuel-based transport emits CO_2 and electric transport raises the issue of electricity production and its environmental impacts. Conversely, roads, railways, vehicles and ways of travelling can be impacted by extreme climate events, such as floods, storms, thawing permafrost and melting asphalt. This second aspect of the relationship between climate change and human mobility is rarely explored, even within the scholarship on 'climate mobility'. Focusing on Central Asia, this chapter presents the specificities of the region regarding the environment–mobilities nexus and highlights the adverse impacts of climate-related mobility disruptions for the populations of the region. The chapter is based on the author's fieldwork in Central Asia, particularly in Tajikistan, and on press articles and scientific literature on the topic. It discusses the complex relationship between mobilities and climate change in Central Asia, addresses the interconnection between climate justice and mobility justice and provides policy recommendations to promote sustainable mobilities and reduce mobility dependence in the region.

Keywords Mobilities · Accessibility · Infrastructure · Climate change · Central Asia

S. Blondin (✉)
University of Neuchâtel, Neuchâtel, Switzerland
e-mail: suzy.blondin@unine.ch

© The Author(s) 2023
R. Sabyrbekov et al. (eds.), *Climate Change in Central Asia*,
SpringerBriefs in Climate Studies,
https://doi.org/10.1007/978-3-031-29831-8_9

1 Introduction: Climate Mobility Studies and Current Debates

The way that climate change sets people on the move has been increasingly studied and discussed in recent years in the context of 'climate-induced migration' or 'environmental migration'.[1] Research in the field has shown the complexity of the climate change–mobility nexus, and has revealed the importance of internal, translocal and short-term mobility for populations impacted by environmental hazards (Boas et al. 2019). The concept of 'climate mobilities' 'pays attention to the multiplicity of climate change-related human mobility (involving immobility, relocation, circular mobility, etc.), its embedding in ongoing patterns and histories of movement, and the material and political conditions under which it takes place' (Boas et al. 2022, 2). On the environment–mobilities relationship, some studies have also highlighted the way everyday mobilities are threatened by environmental hazards, how some places are becoming less accessible in the context of climate change (Olsen et al. 2020; Blondin 2022), and how climate change modifies the way some populations circulate within a city or between rural and urban areas (Tuitjer 2019; Arp et al. 2019); however, these studies remain marginal within the climate mobilities scholarship. Disaster-induced mobility restrictions deserve our attention, since, in areas where mobility disruptions are frequent, the issue of habitability is at stake: reduced accessibility may prevent people from practicing essential mobilities to reach work, education, markets, banking or healthcare facilities, for instance, which may cause or exacerbate socioeconomic issues.

The aim of this chapter is twofold: it seeks to highlight the often-neglected issue of the impacts of climate change on mobility infrastructure and conditions, and to position Central Asia in the scholarly discussions on the relationship between climate change and small-scale mobilities. Based on the scientific literature, press reviews and long-term fieldwork in Tajikistan's Pamir Mountains covering a total of 9 months between 2016 and 2020 (Blondin 2020, 2021), this chapter explores both the effects of human mobility on the environment and the effects of climate change on human mobility. The first section of the chapter discusses the way mobilities impact climate change and the environment in Central Asia through the widespread use of old highly polluting vehicles, the lack of public transport in some areas and ongoing large-scale road-building projects. In the second section, the issue is approached from another angle: the impacts of climate change on mobilities are examined through the case of floods and extreme heat damaging the mobility infrastructure. The chapter then presents a discussion on the importance of studying small-scale and everyday mobilities in the context of climate change, considering the interlinkages between climate justice and mobility justice. The conclusion offers policy recommendations to address the issues highlighted in the chapter, focusing both on promoting sustainable mobilities and on reducing mobility dependence in the region.

---

[1] See climig.com, a comprehensive database on the topic.

2 How Mobilities Impact Climate Change in Central Asia

Multiple studies have provided evidence of the impacts of transport and mobility on climate change globally (see Sheller 2018 for a review). Fossil fuel-based transport causes CO_2 emissions directly, while electric transport raises issues relating to emissions from the production of electricity and the environmental aspects of batteries. Moreover, mobility infrastructure itself impacts the environment since it occupies land and generates air, noise, and light pollution. This section discusses these issues in the context of Central Asia, through the lens of old polluting vehicles, the lack of public transport, and the ecological impacts of new/projected road infrastructure in the region.

2.1 Old Vehicles and Pollution

Increasingly, studies are raising the issue of poor air quality in Central Asian cities. Bishkek, Almaty and Dushanbe have been highlighted as cities that expose their residents to 'higher-than recommended levels of air pollution'.[2] As Sabyrbekov and Overland (2020, 3) note, for instance, 'on some days in 2019, Bishkek had the highest levels of air pollution in the world' (see also OECD 2019). The transport sector is often considered particularly polluting, especially given that in Central Asia a large number of vehicles exceed the norms for the emission of harmful substances into the atmosphere. Central Asian countries do not have special standards for old emitting cars. Nasritdinov (2021, 188) clearly sums this up, writing about air pollution in Kyrgyzstan:

> [The number of cars] has been increasing very steadily too as Kyrgyzstan was joining the Eurasian Economic Union and there were multiple speculations about the increasing tariffs on imported cars. Most cars imported to Kyrgyzstan are second hand and quite old. They do not have proper emission filtering technologies. The solution to cars would be a proper public transport, but that is in crisis too in Bishkek. The city has a very small stock of buses and trolley-buses. Instead, the majority of residents ride in marshrutkas (mini-buses), which themselves pollute the air more than any other vehicles, but they are also often very inconvenient and crowded to ride. So, for many who can afford, cars are a preferred option.

Throughout Central Asia, the use of private cars has greatly increased since the collapse of the Soviet Union and its public transport system. In places where public (trolley-)buses have stopped operating, *marshrutkas*, or minibuses, have emerged as a bottom-up solution to the lack of public transport (see Rekhviashvili and Sgibnev 2020). While *marshrutkas* offer flexibility (they operate along a variety of routes and users can usually stop wherever they want, not only at formal bus stops) and an important source of income for many households, they contribute significantly to the high levels of air pollution in Central Asian cities. Often, old or vulnerable

---

[2] https://blogs.worldbank.org/europeandcentralasia/five-steps-for-cleaner-air-in-central-asia.

vehicles[3] emitting high levels of carbon are used as *marshrutkas*. In many regions throughout Central Asia, the distribution of buses, trams or metros is not sufficient to limit the use of mini- and microbuses and their adverse effects on air quality. While *marshrutkas* usually complement regular public transport, they constitute the only shared transport option in some regions, especially rural ones, such as Tajikistan's Viloyati Mukhtori Kuhistoni Badakhshon (VMKB) region (which extends across half of Tajikistan's total territory).

Even in cities where buses or tramways operate well, the issue of fares also plays an important role (Asia Plus 2021). Public transport is too expensive for a large part of the population, who have no other option than to use the cheaper *marshrutkas*, even though they are often viewed as uncomfortable or as reinforcing gender inequality and social exclusion since they expose passengers to harassment, threats or theft which may prompt them to reduce certain forms of mobilities (Nasritdinov 2021; Turdalieva and Edling 2018). In some areas or at some hours, taxis (i.e. private cars rather than minibuses) are the only motorised option available, and also contribute to high carbon emissions and poor air quality due to their old age and lack of maintenance, and because they carry fewer passengers than *marshrutkas*.

While some transport experts (or development banks, such as the Asian Development Bank in Tajikistan)[4] advocate for the use of electric vehicles in order to limit or eliminate carbon emissions, this 'solution' would not seem suitable for Central Asian countries in the near future since power shortages are still frequent (RFERL 2021), and strongly impact large segments of the population, especially in Tajikistan and Kyrgyzstan. Turning to electric vehicles would raise serious issues in terms of electricity production and distribution in these countries. In addition, it is important to note that, even while the use of electric vehicles might help to reduce carbon emissions, the production and recycling of batteries remain problematic in terms of environmental impacts and sustainability.

2.2 *The Ecological Impacts of Mobility Infrastructure*

While vehicles are central when thinking about the links between mobilities and climate change, we should not forget about the impacts of the mobility infrastructure, such as roads, railways, or airports, itself. Such infrastructure provokes air, noise and light pollution that may have profound impacts on the environment. In Central Asia today, road-building projects are numerous, and many are large scale. The Chinese government's Belt and Road Initiative (BRI) is probably the most talked about project in this domain (Vakulchuk and Overland 2019). While building or renovating roads in Central Asia seems crucial because some areas still suffer from road infrastructure decay, the adverse environmental impacts of road building should be noted. In the case

---

[3] For instance, the Chinese 'Tangen' minibuses in Khorog, Tajikistan, which need almost constant maintenance given their vulnerability to poor road conditions and frequent hazards.

[4] https://development.asia/insight/how-electric-vehicles-can-make-tajikistan-emissions-free.

of large-scale Chinese-led road projects, the environmental impacts of infrastructure projects are often overlooked. As Coenen and colleagues (2021, 4) observe:

> Protecting the environment while fostering economic development under the BRI will be challenging, as the initiative traverses a diverse range of fragile environments. Biophysical conditions range from forests and steppes in Russia, to ice, snow, and permafrost across the Tibetan Plateau, and tropical rainforests in Malaysia.

Building roads, railways or harbours have negative impacts on ecosystems, biodiversity and wildlife, and such infrastructure may encourage new human settlements, which would occupy more land and likely cause deforestation and/or aridification. In addition, the construction of mobility infrastructure itself results in an increase in CO_2 emissions and 'accelerate[s] extraction of natural resources, such as water, sand, and ferrous metal ores in countries along the BRI' (Coenen et al. 2021, 5). Since environmental regulations are getting harsher in China, countries like Tajikistan are also becoming 'pollution havens' where cement is increasingly produced (Coenen et al. 2021).[5]

Within cities as well, new infrastructure is often built at the expense of local biodiversity. For instance, in Bishkek, activists have recently spoken up about the cutting of old urban trees for the sake of road construction.[6] In addition to carbon emissions and loss of biodiversity, the cutting of trees to build infrastructure raises the issue of the adaptation of cities to global warming given that the presence of vegetation is often presented as a solution to mitigate the effects of extreme heat waves within cities (a phenomenon known as the 'urban heat island', see for instance Hiemstra et al. 2017). Thus, the elimination of green areas could have heavy consequences in Central Asian cities where temperatures regularly reach extremes. The presence of green areas in cities also has a crucial impact on urban dwellers wellbeing and connectedness to the biophysical world (Phillips and Atchison 2020).

In sum, in Central Asia, transport has a high environmental cost. Although this could be said of every part of the world, Central Asia faces the specific issue of a lack of public transport—especially electric buses or trams—combined with a heavy reliance on old and highly polluting minibuses and private cars. The famous *marshrutkas* of Central Asia are illustrative: they operate to compensate for the lack of public transport and offer a more flexible and often cheaper way to travel, but this is also more polluting and often less comfortable. Large-scale road-building projects in Central Asia also raise the issue of the ecological impact of mobility infrastructure, especially when it requires the cutting of trees, damages arable lands or grazing areas, or is situated in disaster-prone areas. While mobilities highly influence climate change, climate change in turn damages the mobility infrastructure itself and may disrupt mobilities. The next section explores the ways that climate change impacts mobilities in Central Asia.

---

[5] https://chinadialogue.net/en/pollution/9174-china-shifts-polluting-cement-to-tajikistan/.
[6] https://www.opendemocracy.net/en/odr/tree-cutting-and-pollution-in-bishkek/.

3 How Climate Change Impacts Mobilities in Central Asia

In Central Asia, some roads are frequently blocked by floods or mudslides, increasingly so as temperatures rise and glaciers melt, while other roads are buckling under extreme heat. Road closures pose a threat to the communities who depend on these roads to access products, healthcare, and work or educational opportunities (Blondin 2020). Residents of rural areas (still the majority of the Central Asian population)[7] are particularly affected by mobility disruptions since their livelihoods and quality of life depend on rural–urban mobilities. On a wider scale, the vulnerability of the road infrastructure to environmental hazards reduces economic development and international cooperation. Climate change acts as a threat multiplier in regard to such processes (Lim 2016).

3.1 The Effects of Environmental Disasters on Mountain Roads and Rural–Urban Mobilities

Mountainous areas represent a large part of the Central Asian territory, especially in Tajikistan and Kyrgyzstan. A huge portion of the residents of these countries live in mountainous areas and regularly have to travel along roads that are highly vulnerable to rockslides, landslides, avalanches and/or floods. These hazards are not new and have always been part of mountain dwellers' lives. However, climate change has increased the frequency and intensity of these hazards (Hock et al. 2019) or limited their predictability. The melting of glaciers increases the runoff of many rivers, which triggers floods (ibid.). As Zhao et al. (2010) have shown, permafrost is present in many areas throughout Central Asia, and its thawing poses a serious threat to engineering infrastructure, including roads. The growing unpredictability and intensity of precipitation also provoke more rockslides and avalanches, affecting roads and paths.

Related road maintenance requires significant funding and human resources, especially in remote areas, which often does not seem to be the priority for local governments. In Central Asia's mountain areas, residents often have to repair or clear the roads themselves. In Tajikistan's Bartang Valley, for instance, residents often gather to voluntarily 'repair' *their* road with local means. They clear snow and rocks or attempt to elevate the level of the road when it is flooded using rocks, sand, and wood (Blondin 2020). This kind of repair work is often precarious and unsustainable and puts the individuals involved in dangerous situations.

In the face of environmental hazards, roads and paths are not the only things affected. Old vehicles are also highly vulnerable to challenging road conditions and need regular maintenance. In Tajikistan's VMKB region, old four-wheel vehicles—usually bought second-hand and imported from places such as the United Arab

---

[7] https://data.worldbank.org/indicator/SP.URB.TOTL.IN.ZS?locations=Z7.

Emirates, Qatar or Japan, via Tajikistan's capital, Dushanbe—are used as shared 'taxis' in the absence of public transport. Shared car trips in the VMKB involve frequent stops for repairs and sometimes include unexpected overnight stays after cars break down. Given the remoteness of the region from the main markets, purchasing car parts is usually a long and costly process. This definitely puts into question residents' mobility capacities (also called *motility*; see Blondin 2020), their personal safety, and their potential to move back and forth between rural and urban areas. Given the harsh conditions along mountain roads, residents lack adapted mobility options. They often praise the resilience of Soviet vehicles (jeeps and trucks) on flooded and potholed roads.

In Soviet times, many villages in the Bartang Valley were not connected by road and there were no private cars. For residents of those villages, travelling included long walking trips. However, the *Moscow provisioning system* (Mostowlansky 2017) meant that even remote places such as the Bartang Valley were supplied with the most 'essential' products with the help of trucks and/or helicopters. Public bus services operated on the main road of the VMKB region (the M41), and planes would regularly connect the region with Dushanbe. The collapse of this provisioning system has meant that residents of remote valleys have had to find their own mobility options to reach the products and services they need. Even though environmental hazards are not the only reason for mountain dwellers' lack of motility, the fact that 'essential' mobilities can't be realised when roads are closed following hazards raises the issue of the habitability of affected areas. While the residents of Bartang are strongly attached to their valley and most of them aspire to remain (Blondin 2021), they often point out that *their* road represents a serious risk to their livelihoods. Throughout the mountainous regions of Central Asia, the melting of glaciers and permafrost, the increased intensity of floods, and the growing frequency of hazards raise a concern about mountain dwellers' capacities to circulate and to remain in the places they call home.

3.2 The Effects of Extreme Heat on Asphalt and Railways

Away from mountainous areas, extreme heat waves also affect roads and railways. Studies covering different areas of the globe have highlighted the vulnerability of asphalt and railway tracks to extreme temperatures, with extreme heat possibly resulting in 'rutting deformations and cracking of asphaltic roads' (Makkonen et al. 2014, 692). Qiao and colleagues (2020, 342) also explain that 'flexible pavements are particularly vulnerable to extreme high temperatures that can cause a decrease in bitumen viscosity, potentially aggravating rutting (i.e., permanent deformation), roughness, and cracking'. As an example, in the summer of 2021, during the severe heat wave in Canada and the North-Western part of the USA, 'roads buckled as their asphalt cracked amid the heat' (*The Washington Post* 2021a). Closer to Central Asia, a study carried out in Xinjiang, northwest China, has shown that high temperatures and significant daily temperature differences have caused serious asphalt pavement distresses, such as rutting and shoving.

Given that most of Central Asia's territory is characterised by a continental climate, with high summer temperatures getting even higher under the effects of climate change (Reyer et al. 2017), Central Asian roads are highly vulnerable to such processes. In July 2021, an article by Azatlyk Radiosy[8] reported road buckling under temperatures reaching up to 50 °C in Turkmenistan (Azatlyk Radiosy 2021). Summer 2021 was considered the hottest ever recorded in areas such as Central Asia, with temperatures reaching up to 44 °C in Termez, Uzbekistan (*The Washington Post* 2021b). Under the effects of global warming, such heat waves are predicted to be more frequent and more intense, which could definitely provoke mobility infrastructure failures. In addition to impressive rapid-onset hazards such as floods, rockslides and avalanches, more 'discrete' hazards such as heat waves and permafrost thaw affect mobility infrastructure and mobility conditions. These should be of interest for research on the socioeconomic impacts of climate change, especially in a climate-vulnerable region such as Central Asia. The next section discusses and re-politicises the issues of mobility infrastructure failures, accessibility and habitability in the context of climate change.

4 Discussion: When Climate Justice Meets Mobility Justice

Given Central Asia's topographical and climatic specificities and the region's environmental vulnerabilities, there is an urgent need to examine the way mobilities and climate change intersect in the region (Vakulchuk et al. 2022). It is crucial to ensure that populations can dwell and circulate in city suburbs and rural and/or mountainous areas, by improving public transport and the resilience of infrastructure in the face of environmental hazards. Central Asian societies are very mobile both between rural and urban areas, and regionally, between Russia and Central Asia for instance, hence the need for them to enjoy sufficient mobility capital to realise those mobilities. Studies have showed how translocality may represent a strategy in the face of the adverse effects of climate change (Sakdapolrak et al. 2016). Accessing economic and educational opportunities, products, and services, in urban centres is central to the livelihoods of rural populations; their inability to sufficiently circulate could provoke out-migration. This means that inaccessibility may lead to relocation (Blondin 2022), with territorial accessibility being part of living conditions and perceived habitability.

The 'mobility justice' approach developed by Sheller (2018) re-politicises the issues of mobility capitals, (in)accessibility and the capacity to dwell and to circulate. Studying the effects of mobilities on climate change and the effects of climate change on mobilities is a way to connect mobility justice and climate justice. Questions of how best to promote and adopt resilient and sustainable mobilities to ensure the mobilities of those who really need it in the context of climate change remain. In

---

[8] From the Radio Free Europe/Radio Liberty Group.

Central Asian countries, much is yet to be done to increase the population's awareness about the impacts of mobilities on climatic conditions also to increase the mobility capacities of populations that suffer from a lack of accessibility where they live (Blondin 2020).

5 Going Forward: Policy Recommendations

Although issues around the human mobilities–environment nexus are manifold and complex, Central Asian authorities, alongside the local populations, could work towards promoting sustainable mobilities and increasing accessibility while also reducing mobility dependencies in some areas.

5.1 Solutions for More Sustainable Mobilities

Promoting and developing public transport: In Central Asia, more efficient, accessible and affordable public transport systems would help to reduce the number of private cars and old minibuses in an effort to lower pollution levels and to help residents increase their capacity for mobility. In Sheller's mobility justice approach (2018), this corresponds to a 'commoning of mobilities'.

Promoting green mobilities: In Central Asia, the active promotion of green mobilities is lacking. Authorities should prioritise the promotion of walking and cycling, or, potentially, the use of electric vehicles. For instance, authorities should address the factors that discourage cycling, such as 'being female, being a student, being a civil servant, living in a block of flats, and age' (Sabyrbekov and Overland 2020, 11) in order to create bicycle-friendly environments and make cycling more socially acceptable and valued. In Central Asia, NGOs and other organisations are already promoting cycling in active ways, as in the case of 'Car Free Day' street actions and cycling actions in Dushanbe led by the NGO Malenkaya Zemlya.[9]

Increasing the resilience of mobility infrastructure: Although building climate-resilient roads, railways or bridges is challenging and costly, solutions exist to make mobility infrastructure more robust and sustainable (see for instance engineering options to design 'new climate-resilient roads in the Kyrgyz Republic' by Lim [2016]).

Avoiding new polluting road-building projects or mitigating their environmental impacts: Although Central Asia still needs to improve its road infrastructure in many areas, it is crucial to discuss and mitigate the environmental impacts of all road (re)construction projects which threaten biodiversity and residents' wellbeing, and emit CO_2.

---

[9] https://leworld.org/en.

5.2 Solutions to Reduce Mobility

Reducing mobility dependence: Ensuring better access to 'essential' products and services (healthcare, educational opportunities, banking facilities) in rural areas is crucial to alleviating residents' mobility dependence. Provisioning 'remote' areas could relieve residents from the burden of complicated, dangerous, and costly mobilities.

Improving emergency mobility capacities: In the face of disasters and mobility disruptions, it is crucial to help residents leave, circulate or find shelter. Improving emergency mobility capacities in preparation for such events is critical, and includes developing evacuation, maintenance and repair systems. This would help residents deal with disasters from social, environmental, and mobility justice perspectives (Graham and Thrift 2007; Sheller 2018).

References

Arp CD, Whitman MS, Jones BM, Nigro DA, Alexeev VA, Gädeke A, Fritz S, Daanen R, Liljedahl AK, Adams FJ, Gaglioti BV, Grosse G, Heim KC, Beaver JR, Cai L, Engram M, Uher-Koch HR (2019) Ice roads through lake-rich Arctic watersheds: integrating climate uncertainty and freshwater habitat responses into adaptive management. Arct Antarct Alp Res 51(1):9–23

Asia Plus (2021). Public transport fares in Dushanbe expected to rise on average 50 percent next month, October 15. https://asiaplustj.info/en/news/tajikistan/economic/20211015/public-transport-fares-in-dushanbe-expected-to-rise-on-average-50-percent-next-month

Azatlyk Radiosy (2021) В Туркменистане воздух нагревается выше 50 градусов, возникла пылевая буря, July 2. https://rus.azathabar.com/a/31337558.html

Boas I, Farbotko C, Adams H, Sterly H, Bush S, Van der Geest K, Wiegel H, Ashraf H, Baldwin A, Bettini G, Blondin S, de Bruijn M, Durand-Delacre D, Fröhlich C, Gioli G, Guaita L, Hut E, Jarawura XF, Lamers M, Lietaer S, Nash SL, Piguet E, Rothe D, Sakdapolrak P, Smith L, Tripathy Furlong B, Turhan E, Warner J, Zickgraf C, Black R, Hulme M (2019) Climate migration myths. Nat Clim Chang 9(12):901–903

Boas I, Wiegel H, Farbotko C, Warner J, Sheller M (2022) Climate mobilities: migration, im/mobilities and mobility regimes in a changing climate. J Ethn Migr Stud 48(14):3365–3379. https://www.tandfonline.com/doi/full/10.1080/1369183X.2022.2066264

Blondin S (2020) Understanding involuntary immobility in the Bartang Valley of Tajikistan through the prism of motility. Mobilities 15(4):543–558

Blondin S (2021) Staying despite disaster risks: place attachment, voluntary immobility and adaptation in Tajikistan's Pamir Mountains. Geoforum 126:290–301

Blondin S (2022) Let's hit the road! Environmental hazards, materialities, and mobility justice: insights from Tajikistan's Pamirs. J Ethn Migr Stud 48(14):3416–3432. https://www.tandfonline.com/doi/full/10.1080/1369183X.2022.2066261

Coenen J, Bager S, Meyfroidt P, Newig J, Challies E (2021) Environmental governance of China's Belt and Road Initiative. Env Pol Gov 31:3–17. https://doi.org/10.1002/eet.1901

Graham S, Thrift N (2007) Out of order: understanding repair and maintenance. Cult Soc 24(3):1–25

Hiemstra JA, Saaroni H, Amorim JH (2017) The urban heat island: thermal comfort and the role of urban greening. In: Pearlmutter D et al (eds) The urban forest. Future city, vol 7. Springer, Cham. https://doi.org/10.1007/978-3-319-50280-9_2

Hock R, Rasul G, Adler C, Cáceres B, Gruber S, Hirabayashi Y et al (2019) High mountain areas. In: Pörtner HO, Roberts DC, Masson-Delmotte V, Zhai P, Tignor M, Poloczanska E et al (eds) IPCC special report on the ocean and cryosphere in a changing climate. IPCC, Geneva, p 72

Lim S (2016) Designing new climate-resilient roads in the Kyrgyz Republic. Asian Development Blog. https://blogs.adb.org/blog/designing-new-climate-resilient-roads-kyrgyz-republic

Makkonen L, Ylhäisi J, Törnqvist J, Dawson A, Räisänen J (2014) Climate change projections for variables affecting road networks in Europe. Transp Plan Technol 37(8):678–694

Mostowlansky T (2017) Building bridges across the Oxus: language, development, and globalization at the Tajik-Afghan frontier. Int J Sociol Lang 247:49–70

Nasritdinov E (2021) Politics of green development: trees vs. roads. In: Isaacs R, Marat E (eds) Routledge handbook of contemporary Central Asia. Routledge, Abingdon, Oxon, pp 180–190

OECD (2019) Promoting clean urban public transportation and green investment in Kyrgyzstan, green finance and investment. Éditions OECD, Paris

Olsen J, Nenasheva M, Hovelsrud GK (2020) 'Road of life': changing navigation seasons and the adaptation of island communities in the Russian Arctic. Polar Geogr (latest articles): 1–19. https://doi.org/10.1080/1088937X.2020.1826593

Phillips C, Atchison J (2020) Seeing the trees for the (urban) forest: more-than-human geographies and urban greening. Aust Geogr 51(2):155–168. https://doi.org/10.1080/00049182.2018.1505285

Qiao Y, Santos J, Stoner A-M, Flinstch G (2020) Climate change impacts on asphalt road pavement construction and maintenance: an economic life cycle assessment of adaptation measures in the State of Virginia, United States. J Ind Ecol 24:342–355. https://doi.org/10.1111/jiec.12936

Rekhviashvili L, Sgibnev W (2020) Theorising informality and social embeddedness for the study of informal transport. Lessons from the marshrutka mobility phenomenon. J Transp Geogr 88:102386

Reyer C, Otto IM, Adams S et al (2017) Climate change impacts in Central Asia and their implications for development. Reg Environ Change 17:1639–1650. https://doi.org/10.1007/s10113-015-0893-z

RFERL (2021). The curious case of Central Asia's severe electricity shortages. By Bruce Pannier, November 16. https://www.rferl.org/a/central-asia-severe-electricity-shortages/31564293.html

Sabyrbekov R, Overland I (2020) Why choose to cycle in a low-income country? Sustainability 12–18:7775

Sakdapolrak P, Naruchaikusol S, Ober K, Peth S, Porst L, Rockenbauch T, Tolo V (2016) Migration in a changing climate. Towards a translocal social resilience approach. DIE ERDE – J Geogr Soc Berl 147(2):81–94. https://doi.org/10.12854/erde-147-6

Sheller M (2018) Mobility justice: the politics of movement in an age of extremes. Verso, London

Tuitjer L (2019) Bangkok flooded: re (assembling) disaster mobility. Mobilities 14(5):648–664

Turdalieva C, Edling C (2018) Women's mobility and 'transport-related social exclusion' in Bishkek. Mobilities 13(4):535–550

Vakulchuk R, Overland I (2019) China's Belt and Road Initiative through the Lens of Central Asia. In: Cheung FM, Hong Y-y (eds) Regional connection under the Belt and Road Initiative. The prospects for economic and financial cooperation. Routledge, London, pp 115–133.

Vakulchuk R, Daloz AS, Overland I, Sagbakken HF, Standal K (2022) A void in Central Asia research: climate change. Cent Asian Surv: 1–20. https://doi.org/10.1080/02634937.2022.2059447

The Washington Post (2021a) It's the climate change, stupid. By Ishaan Tharoor, June 30. https://www.washingtonpost.com/world/2021/06/30/climate-change-heat-politics/

The Washington Post (2021b) Record heat bakes Middle East as temperatures top 125 degrees. By Matthew Cappucci, June 7. https://www.washingtonpost.com/weather/2021/06/07/record-june-heat-wave-middle-east/

Zhao L, Wu Q, Marchenko S, Sharkhuu N (2010) Thermal state of permafrost and active layer in Central Asia during the international polar year. Permafr Periglac Process 21:198–207. https://doi.org/10.1002/ppp.688

Suzy Blondin received her PhD in geography from the University of Neuchâtel, Switzerland, in 2021. Her doctoral dissertation examines (im)mobilities to, from and within Tajikistan's Bartang Valley and the way rural–urban mobilities help the Bartangis to remain and to preserve intimate bonds with their Valley despite environmental risks and economic vulnerabilities. Her research also brings forward the issue of involuntary immobility caused by frequent hazard-related road closures and low motilities, which threatens local livelihoods. Her work has been published in Mobilities, Geoforum and the Central Asian Survey.

Open Access This chapter is licensed under the terms of the Creative Commons Attribution 4.0 International License (http://creativecommons.org/licenses/by/4.0/), which permits use, sharing, adaptation, distribution and reproduction in any medium or format, as long as you give appropriate credit to the original author(s) and the source, provide a link to the Creative Commons license and indicate if changes were made.

The images or other third party material in this chapter are included in the chapter's Creative Commons license, unless indicated otherwise in a credit line to the material. If material is not included in the chapter's Creative Commons license and your intended use is not permitted by statutory regulation or exceeds the permitted use, you will need to obtain permission directly from the copyright holder.

A Gendered Approach to Understanding Climate Change Impacts in Rural Kyrgyzstan

Karina Standal, Anne Sophie Daloz, and Elena Kim

Abstract This chapter explores climate change impacts and the related experiences and realities of local women in rural Kyrgyzstan by combining research on the physical impacts of climate change in the Central Asian region with an analysis of ethnographic accounts of local people's farming and energy-use practices. Our analysis reveals how interlinked material, social and cultural realities of local communities manifest in social differentiation that enables or limits women's capacities to cope with climate change and engage in adaptation practices. The post-Soviet period has diminished rural women's access to social protection and economic opportunities while reinforcing patriarchal gender norms, depriving women of land ownership rights and decision-making power over strategic life decisions.

Keywords Gender · Climate change · Agriculture · Energy · Contextual vulnerability · Kyrgyzstan

1 Introduction

The gender and climate change nexus has gained increasing attention from global development institutions and the research community, with the recognition that understanding individuals' positions in society and the political economy impact their capacity to adapt to a changing climate. The academic and practitioner literature has confirmed women's limited opportunities to engage in climate adaptation practices due to their insecure rights to land, limited access to assets and resources, lack of participation in decision-making and missed educational opportunities (Perkins and

---

K. Standal (✉) · A. S. Daloz
CICERO, Oslo, Norway
e-mail: karina.standal@cicero.oslo.no

A. S. Daloz
e-mail: anne.sophie.daloz@cicero.oslo.no

E. Kim
American University of Central Asia, Bishkek, Kyrgyzstan

Osman 2021; Eastin 2018; Nyasimi and Huyer 2017). However, the general literature on climate change impacts, mitigation and adaptation tends to have a technical and economic focus, where gender is assumed to be irrelevant (Perkins and Osman 2021). Some emerging literature focuses specifically on the interlinkages between gender and climate change (e.g. Holmelin 2019; Eastin 2018; Goh 2012). This literature provides an in-depth understanding into the ways in which climate change and pre-existing economic, social and cultural pressures are producing and reproducing social and economic inequality. A major challenge of this literature is that, while the manifestations of gendered vulnerability to climate change are well described, the complex and interlinked factors that give rise to them cannot easily generalised, and cause and effect are not clearly delineated.

This chapter provides a comprehensive and empirically based overview of climate change impacts in rural Kyrgyzstan and explores women's experiences as they adapt to new realities. Valuing local voices (DeVault and Gross 2012) and women's own experiences, needs and perceptions, allows us to critically engage with the implicit structures of oppression and domination in relation to climate change impact and adaptation. We draw our findings from an interdisciplinary approach that combines the analysis of existing regional research on the physical impacts of climate change and ethnographic accounts of rural women's everyday farming and energy-related practices. Both farming and energy use have been identified as being vulnerable to climate change and climate measures globally (Masson-Delmotte et al. 2018). At the local level, they are indispensable dimensions of rural women's everyday life, identity and wellbeing (Kim and Standal 2019; Kim and Ukueva 2017).

Our analysis focuses on contextual vulnerability, whereby climate change impacts are understood in connection with pre-existing pressures, such as migration, declining social protection, and discriminatory cultural gender norms (Goodrich et al. 2019). Further, we draw on the gender and political economy approach, which explore systematic social oppression based on gender, class, ethnicity, religion and age from a materialist perspective (Enloe 2013). This approach highlights the (lacking) social value of women's work and pays particular attention to reproductive work (social reproduction) in society (Perkins and Osman 2021; Enloe 2013). Global trends, such as the feminisation of labour, where women often work within low paid and informal arrangements (Kabeer 2016) and the 'care crisis' in capitalist economies (Fraser 2017), indicate the continued exploitation of women through the reproduction of the conditions of patriarchal structures, where women are associated with the private and unproductive sphere and men the public and productive sphere. This is also reflected in the way that livelihood and care work perspectives and related services continue to be neglected and poorly understood in climate and energy policymaking (Perkins and Osman 2021; Standal et al. 2018), despite overwhelming evidence that reproductive work to create, educate, feed and provide care for the labour force is vital for economic activity, as well as significantly impactful on greenhouse gas (GHG) emissions (Perkins and Osman 2021; Fraser 2017). Moreover, climate governance effectiveness depends on heterogenous participation, especially the inclusion of vulnerable groups (Perkins and Osman 2021).

This approach is especially relevant to the context of Kyrgyzstan, where the post-Soviet transition resulted in women's diminished access to state social protection. Following Kyrgyzstan's independence in 1991, the country introduced a market economy through a 'shock therapy' method that severely depleted social services and reduced economic opportunities and employment. This transition resulted in decreased political participation by women and the reinforcement of patriarchal gender norms inherent in the new post-colonial national ideology (Kim and Karioris 2021). For many rural women, this meant being deprived of their land ownership rights; limited access to financing, markets, training and networking; and being deprived of decision-making power, even about strategic life decisions, such as who and when to marry (Kim and Standal 2019).

2 Gender and Climate Change Impacts on the Agricultural and Energy Sectors in Kyrgyzstan

Kyrgyzstan is an exceedingly mountainous country, which, like the rest of Central Asia (see Daloz, this volume), is highly exposed to climate change (Vakulchuk et al. 2022). Between 1960 and 2010, the annual average temperatures increased by approximately 1.2 °C (WBG and ADB 2021), causing multiple effects, including the melting of glaciers (Barandun et al. 2018; Gan et al. 2014). By the 2050s, the Coupled Model Intercomparison Project Phase 6 (CMIP6) models project 5.3 °C of warming for Kyrgyzstan under the highest emission pathway (SSP5-8.58.5), a rise faster than the global average. Under all emissions pathways, extreme temperatures are projected to increase, enhancing heat stress, especially in lowland areas. The projections for precipitation are more uncertain, as climate models have difficulty in representing this region owing to its complex topography. However, some climate models predict an increase in extreme precipitation—a cause of concern as parts of Kyrgyzstan are already increasingly exposed to floods, landslides and glacial lake outbursts (Chandonnet et al. 2016; Havenith et al. 2015). A lack of precipitation may also become an issue for Kyrgyzstan in the future, as Central Asia is predicted to become one of the regions most affected by meteorological droughts (Naumann et al. 2018; Vakulchuk et al. 2022).

About 65% of the population in Kyrgyzstan lives in rural areas, most of which are mountainous (Murzakulova 2020). However, the agricultural sector's contribution to GDP has drastically decreased, dropping from 44% in the 1990s to 12% in 2017 (Mogilevskii et al. 2017). Rural unemployment, poverty, dysfunctional social protection systems, and massive out-migration all characterise the wider background against which our study takes place (Murzakulova 2020; Tilekeyev et al. 2019; Mogilevskii et al. 2017). Climate projections present imminent threats to rural livelihoods in Kyrgyzstan, especially for households that rely on farming. Farming at high altitudes is challenged by early frosts, long winters, spring floods, droughts, soil salinisation and risk of natural hazards (Mogilevskii et al. 2017; Bobojonov and

Aw-Hassan 2014). Livestock farming is the mainstay of the agricultural economy and enjoys the most investment (Murzakulova 2020; Schoch et al. 2010). Kyrgyz families rely on livestock for subsistence, emergency funds, and cultural identity as herders. This sub-sector depends on the quality of pastures and the animals' health, both of which are vulnerable to environmental change. Drought spells and a general decline in precipitation in the region have also led to crop loss and food vulnerability for poor households, as well as an increase in the price of agricultural produce (Bobojonov and Aw-Hassan 2014; Lioubimtseva and Henebry 2009). Moreover, irrigation, drainage networks and other Soviet-era agricultural infrastructure are weakening, owing to a lack of investment and policy prioritisation (Mogilevskii et al. 2017; Sagynbekova 2017). In this context, climate change impacts, such as an increase in heat stress, could enhance income unpredictability (Bobojonov and Aw-Hassan 2014). Seed sterilisation (due to heat) and extreme weather may have adverse impacts on yields, endangering livelihoods and low-income households, with small-holder and subsistence farmers being most at risk.

The energy sector in Kyrgyzstan has been a source of domestic and cross-border contestation. About 90% of electricity generation in Kyrgyzstan is from hydropower, and the country has been struggling to meet domestic demand owing to water shortages, resulting in public protests against the government. The year 2021 was marked as an energy crisis year as water basins were depleted. Some rural areas have 'cold spots' in the electricity network, which are particularly vulnerable to breakdowns because they need an upgrade or because they are vulnerable to extreme weather events (Kim and Standal 2019). Climate change will result in more extreme events and changes in runoff, precipitation and the melting of glaciers (Gan et al. 2014), leading to further water shortages. At the 2021 United Nations Climate Change Conference (COP26) in Glasgow, the Kyrgyz government restated its plans for increasing the share of hydropower in the overall energy mix, but this requires considerable investment and support that may be difficult to secure. The Kyrgyz government has also pledged to reduce greenhouse gas emissions from fossil fuel use. For rural families, energy fuels such as coal, wood and dung are essential for subsistence (as illustrated in our second case study below). A balance between water demand for energy versus agriculture will also need to be found. In 2021, disputes over irrigation water triggered armed clashes along the border between Kyrgyzstan and Tajikistan, resulting in the death of more than 40 people (Helf 2021).

In Kyrgyzstan, work migration (within the country and overseas) is considered to be an important adaptation strategy to minimise the risks and uncertainty associated with environmental change affecting agricultural income (aggravated by a lack of irrigation schemes), though this trend is formed by complex and intertwined factors (Blondin 2019; Sagynbekova 2017; Chandonnet et al. 2016). Remittances have become an economic mainstay of rural families, but these economic resources are not invested in supporting innovation in climate-smart farming practices (Mogilevskii et al. 2017; Sagynbekova 2017). Rather, migration weakens community-based natural resource management institutions such as pasture committees and water user associations (Murzakulova 2020) and puts pressure on the remaining population (mostly women) to continue farming activities and uphold family ties to ancestral land (Kim and Standal 2019). Women are more likely to engage in small-holding or subsistence agriculture, and they have less agency to implement new technologies and influence consumer and market preferences (Kim et al. 2018; Kim and Ukueva 2017). Access to resources, such as land, water, energy, credit, income and supportive institutional networks, is socially differentiated by gender, class and ethnicity (Sealise and Undeland 2016). At a global level, research shows that male out-migration can increase women's decision-making in the family as they take charge of the household and farming activities, but it simultaneously increases their work burden and insecurity (Alston 2021; Holmelin 2019). Domestic violence is also widespread in Kyrgyzstan (Kim and Karioris 2021), severely affecting women's contextual vulnerability. Studies from the global South find that domestic violence increases after extreme weather events that reduce agricultural output (Eastin 2021; Caridade et al. 2022). All of this adds to women's contextual vulnerability resulting from climate change and associated risks.

Below, we present two ethnographic case studies from rural Kyrgyzstan. The first draws upon Kim and Ukueva's (2017) case study illustrating how climate change transforms women's access to value chains, while the second, based on Kim and Standal's (2019) case study, illustrates women's struggles around energy access and climate change. Both studies are based on already published data, with additional previously unpublished quotes from respondents.

3 Adapting to Uncertainty: Women's Marginalisation in Rural Issyk-Kul Farming

Issyk-Kul Lake in eastern Kyrgyzstan is a major tourist attraction but suffers from a lack of regional formal employment opportunities and social provisions. Local families rely on diverse income sources, including community-based tourism, crop agriculture and animal husbandry. Women cultivate crops to generate food and income, and they accumulate unique knowledge of local crop species (Kim and Ukueva 2017). Animal husbandry, though practised by both women and men, is considered to be a male domain. However, many men migrate abroad or to urban areas of the country in search of cash earnings (Sagynbekova 2016).

The fertile soil of the area around Issyk-Kul has made it conducive for fruit and vegetable cultivation and many rural women grow apricots as a main source of income. One respondent stated that her family had 'lived by apricots' for the last 15 years and the income was enough for them to 'survive in the wintertime'. These women take several measures to increase and safeguard their apricot production, such as using a mix of organic and mineral fertilisers (animal manure and saltpetre) and insecticides, whitewashing every tree, and prayers: 'We pray and ask for good weather, no hail, no storms'. Their situation is undermined by the lack of social value that is attributed to their labour and overall gender relations in their community.

The women use so-called 'heavy trucks' to sell their produce. This refers to the intermediary purchasers who arrive in the village with large trucks to buy the produce in large quantities. These middlemen, all Kyrgyz men from nearby towns, arrive a few times per month during the harvest season. They establish the purchasing price; the women have little say in negotiating prices for fear that if they wait for too long, the produce will become rotten. These purchasers also determine the scheduling of their arrivals, controlling thereby much of the organisation of the apricot sales. Communicative transactions among the middlemen and women are typically one-directional, with the unreciprocated flow of information going towards the villagers. In this arrangement, women are marginal to the market and the supply chain and, as a result, obtain only a small fraction of the real market value of their produce. The women go along with the 'heavy truck' process as their only marketing channel for practical reasons, including the fact that it is predictable, even if it is unfair.

Despite their significant effort, the women's marginal position is exacerbated further by the effects of climate change. The 'predictability' of their interactions with the middlemen has become disrupted due to unexpected increases in temperature. In 2015, following a heatwave and the much earlier onset of the warm season, the apricots ripened before the scheduled arrival of the heavy trucks. By the time the middlemen arrived, most of the produce had gone bad. As a result, the women suffered a considerable loss, both in income and socially. Some of the women tried to use their kitchen gardens to grow staple crops such as potatoes and cabbage to increase their food security and income. A number of women gathered wild sea buckthorn berries for sale. Several also made use of the apricots for their own and families' consumption by making apricot juice that could replace drinks such as Coca-Cola and Fanta. However, these activities did not allow them to compete with large-scale suppliers, and the financial loss forced the families to sell livestock, which depleted their source of protein and emergency funds.

The 2015 events undermined women's financial self-sufficiency and further diminished their limited negotiating power. In most households, the oldest men are in charge of financial and budgetary decision-making, even if both women and men contribute to the family budget. The income produced from the apricot sales is seen as the women's 'personal money', which they are free to spend. They had enjoyed independence procuring items such as 'warm clothes for children, coal and firewood for the house'. Not having their 'personal money' fund replenished destabilised women's sense of autonomy and household wellbeing. For the older generation of women, it was an especially important dimension of their life quality. They derived social and psychological empowerment from the symbolic meaning found in being 'good Kyrgyz grandmothers' who take care of their gardens, enjoy, provide their children living in the cities with 'ecologically clean' fruits and vegetables, and look after their grandchildren outside of the 'smoky cities'. The income loss they experienced following the missed apricot sales in 2015 limited their ability to be good grandmothers who 'pamper' their grandchildren and the associated social and cultural capital.

This case study demonstrates the imminent risk that rural farmers will face as new weather events repeatedly inflict detrimental effects on their produce

(Climate Risk Profile 2021; Masson-Delmotte et al. 2018) or, as in the situation described here, on the sales of the produce. The study also shows women's resilience and creativity in adapting to new situations. Structural gender inequality is likely to increase with climate change. Certainly, both women and men will be affected by climate change, but the gender-differentiated contextual vulnerability means that women will be impacted differently than men owing to the gender division of labour, male control of supply chains and finances, and the associated marginalisation and disempowerment of women.

4 Energy Struggles: Energy and Women's Care Work in Rural Naryn

Naryn, in Central Kyrgyzstan, is one of the country's poorest regions, with little industry and few non-farm work opportunities, especially for women. The region is located 2000–2500 m asl with long and cold winters, increased precipitation (snow) and limited farming opportunities beyond livestock husbandry. Most households rely on the work migration of both women and men, who often leave the elderly and children behind to take care of farming activities and care work and uphold the link to ancestral land. Against this backdrop, the electricity supply of Naryn is outdated and prone to breaking down during winter storms, a situation further aggravated by the energy deficit (related to water shortages affecting the hydropower network). As a result, people are compelled to use traditional energy sources, such as animal dung, firewood and coal, for heating and cooking. Access to these resources is socially differentiated, as it involves hard physical labour, financial costs and transport arrangements. In general, the supply of energy is formally a male domain, as men prepare animal dung and gather firewood and organise transport trucks for this purpose; as well as purchasing coal in bulk. Women and poor households are generally marginalised in this process. Women do not handle transport or forest department permits (for logging), and poorer households lack livestock for dung and can only afford to buy coal and wood piecemeal at higher unit costs. In line with other studies (Blondin 2019; Sagynbekova 2017; Chandonnet et al. 2016), several of the women and men interviewed stated that people, especially the younger generation, migrate to 'escape harsh winters and find a better and more comfortable life'. Some women were also reported to migrate as a consequence of forced marriage (bride kidnappings), often accompanied by domestic violence. Moreover, land rights are customarily only given to male relatives and livestock husbandry is a male domain, limiting women's livelihood opportunities and financial security.

The lack of resources to secure livelihoods, and the constant struggle to find sources of energy to keep warm and prepare food, affect the life quality of all households that are not affluent. For women, this also has health effects, as most households use the traditional large brick Pechka stoves to cook, which is time-consuming and labour-intensive, and the stoves emit particles that cause indoor air pollution.

Only the most affluent households have invested in electric hotplates for boiling water or gas ovens for cooking. These households practice so-called 'fuel stacking', whereby modern and traditional cooking technology is used side by side. Cooking traditionally (using a Kazan cooking pot on a Pechka stove) is a way for women to seek increased social and cultural status in the family and community. Cooking family meals is perhaps the energy labour most consistently assigned to women and is closely interlinked with identities based on gender roles of femininity, loving, caring, and hospitality (Standal and Winther 2016). A few women had started small businesses to increase their income and independence, such as a bakery, a mobile phone repair business, and vegetable growing, but the unstable electricity supply severely interrupted their activities (Kim and Standal 2019).

As this case study demonstrates, the Kyrgyz government's failure to prioritise women's needs, together with the deterioration of energy provision, compounds women's vulnerable positions in the present-day household economy. Furthermore, the withdrawal of state-sponsored care for the elderly has placed this responsibility onto women because of the taken-for-granted gender division of labour. In addition, the government's failure to stimulate job creation for the rural population has resulted in massive out-migration, with several children being left in the care of elderly grandparents whose health resources are limited.

Ironically, as a former part of the Soviet Union, Kyrgyzstan was fully electrified during the 1950s (Reid 2005). Rural electrification was prioritised as a way to transform 'backward' and 'unhygienic' communities, and to transform 'subordinated' women (under traditional patriarchies) into active citizens engaged in the productive economy (Reid 2005). Electricity has been linked to opportunities for women's empowerment by simplifying domestic work and supporting economic opportunities (Winther et al. 2017). The gender dimension of this in Kyrgyzstan is revealed in the way that women are marginalised in their agency and in the opportunities available to them to make choices relating to their livelihoods and social care provision. This is associated with how the reproductive economy is assigned a lower value in the masculine-dominated political economy where women's issues are essentially rendered invisible and irrelevant. The social implications of the government's lack of understanding and prioritisation of energy services for rural populations are likely to increase further as climate change puts both the energy sector and rural communities under pressure.

5 Seeing Beyond Energy and Livelihoods: Women's Vulnerabilities in a Changing Climate

This chapter has advanced the idea that the interplay of contextual conditions creates multiple layers of vulnerabilities in accessing the social and material capital necessary to mitigate climate change impacts. Drawing on women's voices, this chapter reveals that gendered structural discrimination in regard to their rights to land, access to income, and access to extension services and basic resources make it difficult for women to negotiate fair conditions in their access to markets, employment and care work. Their wellbeing and struggles are furthermore not understood or prioritised at the policy level even though they bear the bulk of the social reproduction burden. This crisis of care and social protection reproduces and increases inequalities. Some families have the economic and social capital to provide social care (e.g. being able to acquire sufficient energy sources and not migrating away from their children), while others do not. Despite variation in terms of their socio-economic status, age and position within their family and kin group, the devaluation and exploitation of women's work (tending to their family's meals or trying to scramble income by growing apricots) is a consistent factor. As women's adaptive capacity in the face of climate change is entangled with their marginal position in the political economy, their contextual vulnerability will increase as climate change accelerates unless drastic policy measures are implemented.

References

Alston M (2021) Gendered livelihood adjustments in the context of climate-induced disasters. In Eastin J, Dupuy K (eds) Gender, climate change and livelihoods: vulnerabilities and adaptations. Cabi, Oxfordshire, pp 174–184

Barandun M, Huss M, Usubaliev R, Azisov E, Berthier E, Kääb A, Bolch T, Hoelzle M (2018) Multi-decadal mass balance series of three Kyrgyz glaciers inferred from modelling constrained with repeated snow line observations. Cryosphere 12:1899–1919. https://tc.copernicus.org/articles/12/1899/2018/

Blondin S (2019) Environmental migrations in Central Asia: a multifaceted approach to the issue. Cent Asian Surv 38(2):275–292. https://doi.org/10.1080/02634937.2018.1519778Bolch etal.2006

Bobojonov I, Aw-Hassan A (2014) Impacts of climate change on farm income security in Central Asia: an integrated modeling approach. Agr Ecosyst Environ 188:245–255. https://doi.org/10.1016/j.agee.2014.02.033

Caridade SMM, Vidal DG, Dinis MAP (2022) Climate change and gender-based violence: outcomes, challenges and future perspectives. In: Leal Filho W, Vidal DG, Dinis MAP, Dias RC (eds) Sustainable policies and practices in energy, environment and health research. World Sustainability Series. Springer, Cham. https://doi.org/10.1007/978-3-030-86304-3_10

Chandonnet A et al (2016) Environment, climate change and migration in the Kyrgyz Republic. International Organization for Migration (IOM), Bishkek

DeVault M, Gross G (2012) Feminist qualitative interviewing: experience, talk, and knowledge. In: Hesse-Biber S (ed) The handbook of feminist research theory and praxis. Sage, Thousand Oaks

Eastin J (2018) Climate change and gender inequality in developing states. World Dev 107:289–305

Enloe C (2013) Seriously! Investigating crashes and crises as if women really mattered. University of California Press, Berkeley

Fraser N (2017) Crisis of care? On the social reproductive contradictions of Contemporary capitalism. In: Bhattacharya T (ed) Social reproduction theory: remapping class, recentering oppression. Pluto Press, London

Gan et al (2014) Effects of projected climate change on the glacier and runoff generation in the Naryn River Basin, Central Asia. J Hydrol 523:240–251. https://doi.org/10.1016/j.jhydrol.2015.01.057

Goh A (2012) A literature review of the gender-differentiated impacts of climate change on women's and men's assets and well-being in developing countries. CAPRR Working paper 106 International Food Policy Research Institute, Washington DC

Goodrich CG, Udas PB, Larrington-Spencer H (2019) Conceptualizing gendered vulnerability to climate change in the Hindu Kush Himalaya: contextual conditions and drivers of change. Environ Dev 31:9–18. https://doi.org/10.1016/j.envdev.2018.11.003

Havenith HB, Torgoev A, Schlögel R, Braun A, Torgoev I, Ischuk A (2015) Tien Shan geohazards database: landslide susceptibility analysis. Geomorphology 249:32–43. https://doi.org/10.1016/j.geomorph.2015.03.019

Helf G (2021) Border clash between Kyrgyzstan and Tajikistan risks spinning out of control. United States Institute of Peace, May 4. https://www.usip.org/publications/2021/05/border-clash-between-kyrgyzstan-and-tajikistan-risks-spinning-out-control

Holmelin NB (2019) Competing gender norms and social practice in Himalayan farm management. World Dev 122:85–95. https://doi.org/10.1016/j.worlddev.2019.05.018

Kabeer N (2016) Gender equality, economic growth, and women's agency: the "endless variety" and "monotonous similarity" of patriarchal constraints. Fem Econ 22(1):295–321

Kim E, Karioris FG (2021) Bound to be grooms: the imbrication of economy, ecology, and bride kidnapping in Kyrgyzstan. Gend Place Cult 28(11):1627–1648. https://doi.org/10.1080/0966369X.2020.1829561

Kim E, Standal K (2019) Empowered by electricity? The political economy of gender and energy in rural Naryn. Gend Technol Dev 23(1):1–18. https://doi.org/10.1080/09718524.2019.1596558

Kim E, Ukueva N (2017) Gender, poverty and environment in rural Kyrgyzstan: issues of natural resource management, biodiversity and environmental degradation. UNDP Kyrgyzstan, Bishkek

Kim E et al (2018) The making of empowered women: exploring gender and development practice in Kyrgyzstan. Cent Asian Surv 37(2):228–246. https://doi.org/10.1080/02634937.2018.1450222

Lioubimtseva E, Henebry GM (2009) Climate and environmental change in arid Central Asia: impacts, vulnerability, and adaptations. J Arid Environ 73:963–977. https://doi.org/10.1016/j.jaridenv.2009.04.022

Masson-Delmotte V, Zhai P, Pörtner H-O, Roberts D, Skea J, Shukla PR, Pirani A, Moufouma-Okia W, Péan C, Pidcock R, Connors S, Matthews JBR, Chen Y, Zhou X, Gomis MI, Lonnoy E, Maycock T, Tignor M, Waterfield T (eds) (2018) Global warming of 1.5°C. An IPCC Special Report on the impacts of global warming of 1.5°C above pre-industrial levels and related global greenhouse gas emission pathways, in the context of strengthening the global response to the threat of climate change, sustainable development, and efforts to eradicate poverty. IPCC.

Mogilevskii R et al (2017) The outcomes of 25 years of agricultural reforms in Kyrgyzstan. No. 162. Discussion Paper, Leibniz Institute of Agricultural Development in Transition Economies

Murzakulova A (2020) Rural migration in Kyrgyzstan: drivers, impact and governance. Research paper # 7. https://ucentralasia.org/media/pdcnvzpm/uca-msri-researchpaper-7eng.pdf

Naumann G, Alfieri L, Wyser K, Mentaschi L, Betts RA, Carrao H, Spinoni J, Vogt J, Feyen L (2018) Global changes in drought conditions under different levels of warming. Geophys Res Lett 45(7):3285–3296. https://doi.org/10.1002/2017GL076521

Nyasimi M, Huyer S (2017) Closing the gender gap in agriculture under climate change. Agric Dev 30:37–40

Perkins PE, Osman B (2021) Bringing women's livelihood and care perspectives into climate decision-making. In: Eastin J, Dupuy K (eds) Gender, climate change and livelihoods: vulnerabilities and adaptations. Cabi, Wellington

Reid SE (2005) The Khrushchev kitchen: domesticating the scientific-technological revolution. J Contemp Hist 40(2):289–316. https://doi.org/10.1177/0022009405051554

Sagynbekova L (2016) The impact of international migration: process and contemporary trends in Kyrgyzstan. Springer, Cham

Sagynbekova L (2017) Environment, rural livelihoods, and labor migration: a case study in central Kyrgyzstan. Mt Res Dev 37(4):456–463. https://doi.org/10.1659/MRD-JOURNAL-D-17-00029.1

Schoch N, Steimann B, Thieme S (2010) Migration and animal husbandry: competing or complementary livelihood strategies. Evidence from Kyrgyzstan. Nat Resour Forum 34(3):211–212. https://doi.org/10.1111/j.1477-8947.2010.01306.x

Sealise E, Undeland A (2016) Kyrgyz Republic: women and community pasture management. https://www.landesa.org/wp-content/uploads/2016-Best-Practices-Case-Kyrgyzstan.pdf.

Standal K, Winther T (2016) Empowerment through energy? Impact of electricity on care work practices and gender relations. Forum Dev Stud 43(1):27–45. https://doi.org/10.1080/08039410.2015.1134642

Standal K, Winther T, Danielsen K (2018) Energy politics and gender. In: Hancock K, Allison J (eds) Oxford handbook of energy politics. Oxford University Press, Oxford

Tilekeyev K et al (2019) Analysis of youth labor market trends in Kyrgyzstan. Working Paper 52Institute of Public Policy and Administration

Vakulchuk R, Daloz AS, Overland I, Sagbakken HF, Standal K (2022) A void in Central Asia research: climate change. Cent Asian Surv: 1–20. https://www.tandfonline.com/doi/full/10.1080/02634937.2022.2059447

WBG and ADB (2021) Climate risk country profile: Kyrgyz Republic. World Bank Group, Washington, DC and Asian Development Bank, Manila. https://openknowledge.worldbank.org/handle/10986/36377

Winther T, Matinga MN, Ulsrud K, Standal K (2017) Women's empowerment through electricity access: scoping study and proposal for a framework of analysis. J Dev Effect 9(3):389–417. https://doi.org/10.1080/19439342.2017.1343368

Karina Standal is a human geographer working on energy, gender and development. Her main research interests are within the fields of political and development geography, with a specific focus on decentralized renewable energy systems, electrification, sustainable energy consumption and gender relations. Her geographical focus covers Norway, South Asia and Central Asia. She has fieldwork experience from Norway, India, Afghanistan and Kyrgyzstan.

Anne Sophie Daloz is a climate scientist at CICERO, Norway. Her work relies on the analysis of climate data from observations to climate models outputs. She has worked on a variety of regions and processes such as Central Asia or Europe, and tropical cyclones, snowfall, clouds or precipitation. She also has a strong interest in connecting climate sciences to other disciplines.

Elena Kim is a professor of Social Sciences at the American University of Central Asia, Bishkek Kyrgyzstan. Elena's teaching, research and activism focus on intersecting issues of gender, international development, violence, and gender politics in Central Asia. Her publications include chapters in several books and articles including Violence against Women, Journal of Gender Studies, Gender, Technology and Development, Central Asian Survey, Rural Society, and Women and Therapy.

Open Access This chapter is licensed under the terms of the Creative Commons Attribution 4.0 International License (http://creativecommons.org/licenses/by/4.0/), which permits use, sharing, adaptation, distribution and reproduction in any medium or format, as long as you give appropriate credit to the original author(s) and the source, provide a link to the Creative Commons license and indicate if changes were made.

The images or other third party material in this chapter are included in the chapter's Creative Commons license, unless indicated otherwise in a credit line to the material. If material is not included in the chapter's Creative Commons license and your intended use is not permitted by statutory regulation or exceeds the permitted use, you will need to obtain permission directly from the copyright holder.

Climate Change Awareness, Norms and Stakeholders in Central Asia

The Institutionalisation of Environmentalism in Central Asia

Filippo Costa Buranelli

Abstract In 2021, in a largely ignored yet significant step towards regional coordination and convergence, the Central Asian republics took part in the 26th Conference of the Parties in Glasgow as a single entity, speaking with one voice and advocating a joint approach to climate change. Yet, to what extent is Central Asia complying with the norms and rules of environmental governance? Are environmental protection, climate-change mitigation and the push for an energy transition merely a set of shallow practices and rhetoric to signal performative compliance, or are they based on a logic of appropriateness and embedded in a normative understanding of green politics? Drawing on recent scholarship on international society and based on the assumption that environmentalism is now an established institution of the contemporary international order, this chapter considers whether, and in what way, Central Asia has embraced the institution of environmentalism, exploring discourses and practices at the global, regional and local levels. Far from being an exercise in pure theorisation, this can help shape policy engagement from and with the region, allowing us to assess the depth of commitment of these republics and societies in fighting climate change by distinguishing challenges deriving from structural, instrumental or ideological factors.

Keywords Environmentalism · Institutionalisation · Central Asia · Climate change · International norms

1 Introduction

From 31 October to 12 November 2021, the 26th Conference of the Parties to the United Nations (UN) Framework Convention on Climate Change took place in Glasgow, Scotland. Given the dire situation in which the world finds itself with respect to climate change and future predicted environmental disasters, there were

F. Costa Buranelli (✉)
University of St Andrews, St Andrews, Scotland, UK
e-mail: fcb7@st-andrews.ac.uk

great expectations as to what concrete, bold moves could be agreed on at this event. One of the most striking aspects of the conference was that, in a rare display of international multilateralism and unity, the Central Asian republics (Kazakhstan, Kyrgyzstan, Tajikistan, Turkmenistan and Uzbekistan) took part as a single regional group with a unitary voice, sharing proposals and even the same pavilion under the slogan '5 countries, 1 region, 1 voice' (ECIFAS-TJ 2021).

This unity, while surprising because of the continuous underlying tensions that fracture the region, especially regarding border disputes and problems related to water management, is perhaps less surprising given the importance of the environment, both historically and socially, for Central Asia. It is not by chance that in one of the sharpest analyses produced on the relationship between environment and society in Central Asia, the late Shirin Akiner aptly used the concept of 'symbiosis' to describe the importance that the steppe and the oases have played for the Central Asian populations economically, culturally, socially and even religiously (Akiner et al. 2020). After all, the nexus between the environment, development and security is a fundamental factor affecting the regional landscape of the Central Asian republics. In this respect, it is worth remembering that, not without difficulty, the five states have managed to set up the only nuclear-weapon-free zone in the northern hemisphere, in a region surrounded by great nuclear powers, precisely thanks to the way in which environmental and human security have been pitched both at the inter-state level and across civil society in the area, mostly thanks to the push provided by organisations such as the international anti-nuclear movement Nevada-Semipalatinsk (founded in 1989 in Kazakhstan and led by the poet Olzhas Suleimenov) towards fostering a green consciousness and a prototype of green activism in the region.

Drawing on this background, this chapter provides initial answers to the following question: *Has environmentalism been institutionalised in Central Asia?* 'Environmentalism' in this context is defined as a set of principles, discourses, behaviours and norms aimed at protecting the planet and humanity from the effects of climate change and fostering a way of living that is respectful of the environment. The term 'institution' refers to deep and relatively durable social practices which have evolved rather than being designed. Given that an international relations lens is being applied here, these practices must not only be shared by members of international society (i.e. states), but also be recognised by them as legitimate behaviour (Buzan 2004). In this context, institutions are thus about the shared identity of members of international society. They are constitutive of both individual states and international society as a whole in that they define not only the basic character of states but also their patterns of legitimate behaviour in relation to each other, as well as the criteria for membership in international society—thus, they have a regulatory as well as a constitutive dimension.

While some works exist on climate change and Central Asia, especially from a natural sciences perspective (Liu et al. 2020; Yu et al. 2021), the discipline of international relations (IR) is still in its infancy when it comes to assessing the status of environmentalism in the region (Vakulchuk et al. 2022). When an IR prism is adopted to study environmental politics and related concerns in the region, the analysis often reverts to common tropes of security, conflict, the scramble for resources and dynamics reminiscent of the New Great Game narrative (for an exception see Weinthal 2002). Instead, this chapter seeks to advance recent scholarship on norms, rules and legitimacy in (regional) international governance from a Central Asian perspective, and by relying specifically on the concept of 'institution' intends to unpack the socio-structural incentives and constraints that Central Asia faces when it comes to environmentalism. In this respect the chapter may be seen as an advancement in scholarship and a pioneering work, both in terms of *topic* and in terms of *theory* adopted.

In order to understand whether, and to what extent, environmentalism has been established as an institution in Central Asia, I have had to limit the scope of the research. First, in terms of geographical area, this chapter will consider 'Central Asia' to be the five post-Soviet republics of Kazakhstan, Kyrgyzstan, Tajikistan, Turkmenistan and Uzbekistan, although with the awareness that, especially with respect to such a transnational and 'wicked' problem as the environment, borders and boundaries are meaningless (Falkner 2021). This, however, is done for necessity and for coherence with the rest of this volume.

Second, in order to assess the degree of institutionalisation, this essay will consider discourses and practices (the fundamental components of institutions in a sociological sense) at the international, regional and domestic levels. This is done because, if institutionalisation is manifest only in international forums, without sustained processes within the region itself, it would likely amount only to institutional mimicking, that is to say, a display of virtuosity absent implementation. The level of institutionalisation, then, goes hand-in-hand with state machinery, i.e. the development of structures, bodies, legislation, initiatives and other practices that demonstrate a commitment to the principle at the heart of the institution under examination, and the willingness to consolidate it.

Third, as institutions contain a deontic component of appropriateness and conformity, the analysis will also consider the elements of 'rightful conduct', 'necessity' and 'moral need' to comply with environmental norms. This, again, is the advantage of the sociological take on institutions as opposed to a mere cost–benefit analysis.

What my analysis seeks to offer, therefore, is a form of middle-range theorising about environmentalism in Central Asia, which takes into consideration the 'birdseye view' of the international and regional levels, as well as the main institutional markers within states. Conscious that this chapter relies on a state-centric understanding of international relations, what I will not take into account here is the role of civil society activism and bottom-up initiatives, as this is covered by other chapters in this volume. The hope, however, is that this contribution will serve as a useful, if preliminary, overarching framework within which to contextualise and situate the 'greening' of Central Asia.

2 The International Level

At the international level, it is easy to verify the institutionalisation of environmentalism in the Central Asian republics. A good starting point is the signing and ratification of the Paris Agreement, the biggest multilateral climate change related treaty in force at present which aims to bring all nations into a common process to undertake ambitious efforts to combat climate change and adapt to its effects (Paris Agreement, n.d.).

To begin with, all five Central Asian republics have signed and ratified the Paris Agreement, which entered into force in 2016. They have also all undergone at least once an Environmental Performance Review, an important voluntary peer-review mechanism to support member countries of the United Nations Economic Commission for Europe in improving their environmental management and performance (UNECE, n.d.). In terms of other commitments from an international law perspective, the picture is a bit more mixed. In fact, the number of environmental agreements in force, including both multilateral and bilateral documents as well as binding and non-binding, are as follows: 323 for Kazakhstan; 195 for Kyrgyzstan; 160 for Tajikistan; 186 for Turkmenistan and 155 for Uzbekistan (Mitchell 2022).

Although these numbers vary for a variety of reasons, such as the relevance of the treaty for a given country, or simply the fact that 'not all states possess the same capacity to deal with similar environmental problems' (Weinthal 2002, 12), they denote a commitment at the international level to incorporate relevant international environmental legislation within the respective domestic legislative orders. Furthermore, the international dimension of the legitimacy of environmentalism as a standard of conduct, and as a constitutive as well as regulatory practice of post-Cold War international politics, is visible in the number of statements made by Central Asian leaders from all republics since 1991. While for reasons of space, it is impossible to report all speeches in this chapter, three examples may suffice.

Firstly, at the 17th plenary session of the UN General Assembly in 1994, Uzbek Foreign Minister Abdulaziz Kamilov maintained that

> We are grateful to the United Nations and to the specialized agencies concerned with environmental control and with the prevention of global ecological disasters. We are ready to provide all possible assistance in this noble task. (Kamilov 1994)

Secondly, at the 19th special session of the UN General Assembly in 1997, the First President of Kazakhstan Nursultan Nazarbayev proclaimed that

> It is therefore very important to adhere strictly to the principles of the Rio Declaration, ensuring that economic growth takes place only in relationship to processes of social development and environmental security. (Nazarbayev 1997)

And thirdly, and more recently, Kyrgyz President Sadyr Japarov argued in front of the UN General Assembly at its 76th plenary session that

> For three decades, Kyrgyzstan has been an active promoter of the interests of landlocked mountain states in the international arena in order to address the problems of sustainable development and the impact of climate change. (quoted in Osmonalieva 2021)

An analysis of these speeches sheds light on the fact that, not only have the Central Asian republics been legitimising environmentalism as an institution of international society, but they have also been insisting on the role of the UN in spearheading the way in addressing inequalities, dangers and insecurity deriving from climate change and environmental degradation. Moreover, it is not just the UN that is being addressed, but also and especially the complex cosmos of institutions, agencies, donors and epistemic communities that play a role in keeping the environmental spotlight (as well as welcome investments and programmes) focused on Central Asia.

At the international level, the Central Asian states have launched several initiatives over the years aimed at drawing the international community's attention towards environmental issues in the region. These include the 'International Decade for Action: Water for Sustainable Development, 2018–2028', initiated by Tajikistan (Rahmon 2021), and a new draft resolution entitled 'Nature knows no borders: transboundary cooperation is a key factor in the conservation and sustainable use of biodiversity' proposed by Kyrgyzstan (Zheenbekov 2020). These initiatives, while symbolic, have had the effect of legitimising, sustaining and enhancing a 'green discourse' in, about and from Central Asia, which positions the region firmly within the process of institutionalisation of environmental stewardship (Falkner and Buzan 2019). These initiatives also have the merit of creating a financial, normative and bureaucratic conduit between international society and the region. At COP26, this was evident in the words of Zafar Makhmudov, the Executive Director of the Regional Environmental Center for Central Asia. According to him, the joint pavilion provided Central Asian countries 'with **a unique opportunity to demonstrate their investment potential**, their **role in the global climate process**, their perspectives and current **needs** for financing, technologies and expertise' (CARECECO 2021). This leads us to the regional level.

3 The Regional Level

At the regional level—that is, relating to the international relations *between* the Central Asian republics—the institutionalisation of environmentalism started as early as 1992 and continued throughout the 1990s with, first, an inter-ministerial agreement between the newly independent republics (1992), then the agreement on the Aral Sea basin, signed in Qyzylorda, Kazakhstan, which gave birth to the International Fund for the Aral Sea (IFAS), then with the Nukus Declaration also relating to the Aral Sea in 1996 (Uzbekistan) and another region-wide inter-ministerial agreement on the rationalisation of the use of water in 1998. After that initial phase of institutionalisation, though, the early and mid-2000s did not yield many results in terms of environmental cooperation.

More recently, however, things have improved. Under the aegis of the Eurasian Economic Union (EAEU), for example, negotiations are underway at the level of presidents and heads of government of Central Asia to determine and approve the water, energy and food balances (that is to say the sustainable equilibrium of energy, water, and food exchanges between member states) in light of climate change (Masalieva 2022). IFAS aside, there are very few autochthonously institutionalised mechanisms for environmental protection and for mitigating the effects of climate change. There is a passing reference to 'recognising the importance of consolidating efforts' in the climate-change sphere (paragraph 17) within the 2021 Joint Statement of the Heads of State of Central Asia (Akorda 2022) and there is also the Green Bridge Initiative, launched by Kazakhstan in 2010, which has now entered the 2021–2024 phase of programme development. The latter, however, serves as another platform to conduct multi-stakeholder dialogue and activities in the region, as opposed to creating a mechanism for effectively monitoring the progress of the region towards reaching the stated objectives and environmental goals. It is also worth noting that Kazakhstan (the initiator) and Kyrgyzstan are the only two Central Asian countries represented in the initiative, although at COP26 Tajikistan signed an association agreement with Kazakhstan relating to the Green Bridge Initiative (Shayakhmetova 2021).

Keeping in mind the dual aspects of the institutions outlined above, i.e. regulatory and constitutive, it can be said that environmentalism does not play a significant constitutive role in Central Asia either. While countries occasionally look at each other comparing their own domestic situation with that of their neighbours in terms of economic development or political governance, seldom if ever does environmentalism work as a bond, or as a 'standard of good governance', between the Central Asian states. In this respect, therefore, it seems that environmentalism falls under the rubric of sovereign prerogatives and non-interference despite the obvious transnational nature of climate change and environmental degradation.

The institution of environmentalism in Central Asia is kept alive by international donors, organisations and consortia. For example, Central Asia has its own Climate Change Conference (CACCC), which has now been held for four years. CACCC is a continuation of the World Bank's initiative on climate change knowledge and regional information exchange in Central Asia, launched in 2013, and is supported by the Regional Environmental Centre for Central Asia (RECCA-CAREC) jointly as part of the World Bank/IFAS project 'Climate Adaptation and Mitigation Program for the Aral Sea Basin'. It was in fact in the course of one of these conferences, held in Dushanbe in 2021, that the five Central Asian representatives developed the position that was then presented at COP26 in Glasgow, thus showing again the deep 'internationalisation' of the institution of environmentalism, that is, its dependence on the international community. In fact, it is CAREC, in partnership with the Interstate Commission for Sustainable Development (ICSD)[1] which was founded by the five Central Asian heads of state after the Qyzylorda meeting in 1993, that provides the best example of the 'dual-track' institutionalisation of environmentalism between the international and regional levels. The regional statement 'Voice of Central Asia', unanimously adopted by all the Central Asian states and presented at COP26 last November, was developed with the support of RECCA-CAREC in coordination with

---

[1] http://www.mkurca.org/mkur/polozhenie_mkur/.

ICSD, which also advocated for the establishment of the Regional Center for Climate Action Transparency in Central Asia (Statement 'Voice of Central Asia' 2021).

It is in macro-regional frameworks that one would have to look for traces of the institutionalisation of environmentalism. Perhaps not surprisingly, it is the EAEU that, in its treaty, contains most of the references made to the necessity to preserve the environment, thus signalling a rising awareness from the mid-2010s onward of the need to take into account the progressive importance of environmentalism as a norm (EEU Treaty 2015). The treaty contains the term 'environment*' 32 times, of which only 7 occurrences refer to the 'business' or 'investment' environment.[2] In regard to other Russia-led initiatives, the Charter of the Commonwealth of Independent States identifies 'the environment' as an area of cooperation in Articles 4 and 19, whereas nothing is mentioned in the Collective Security Treaty Organisation charter.

Almost specular to the CAREC initiative, is the environmental regionalism promoted by China[3] through the Shanghai Cooperation Organisation (SCO), which addresses climate change and environmental stewardship by 'encouraging efficient regional cooperation in such areas as [...] environment protection' (Art. 1) (SCO Charter 2002). The SCO's framework seems to be the most developed and structured process to address the effects of climate change in the region within its Program of Multilateral Trade and Economic Cooperation and, more importantly, in the SCO Development Strategy 2015–2025, which establishes an explicit link between economic growth and environmental protection.

In the context of the UN-sponsored World Environment Day in 2020, the SCO Secretary-General, Vladimir Norov, reaffirmed the centrality of environmental protection to the SCO cooperation agenda, declaring that 'environmental issues have become one of the main components of economic models for the development of States, including the SCO member states, which have reached to concrete understandings on them' (Agostinis and Urdinez 2021). In this regard, China has been proactive in not simply fostering environmental discourses, norms and strategies with SCO countries (including Central Asia), but also playing a role in rebuilding the Central Asian Power System through the Moinak Hydroelectric Power Plant in south-eastern Kazakhstan and the Nurek Hydropower Plant in Tajikistan with financial support from the Asian Infrastructure Investment Bank and the World Bank, thus promoting an agenda based on the development of green energy.

This shows that, despite the lack of local, indigenous initiatives coming directly from Central Asian states, one can observe that the region is willing to be involved in a broader, transregional polyarchic network of donors, investors, states and companies that is advancing an environmental agenda in the region. In the case of the SCO and China, it is also important to stress the way that environmental norms are not detached from interest-based and geopolitical considerations, especially if the promotion of these norms helps China to advance its agenda in Central Asia

---

[2] https://www.un.org/en/ga/sixth/70/docs/treaty_on_eeu.pdf.

[3] China is the world's largest energy consumer and emitter of CO_2. At the same time, in the quest to diversify and de-carbonize its energy matrix, China has become the world leading investor in renewable energies (Agostinis and Urdinez 2021).

and the Central Asian leaders can extract rents from green projects. Whether or not you consider 'environmentalism' to be a 'liberal' norm, the above shows that non-democratic states and 'illiberal ecologies' can also embrace green agendas and foster the institutionalisation of environmentalism.

4 The Domestic Level

The last step in discussing the institutionalisation of environmentalism in Central Asia is to look at the domestic level to see whether the principles enunciated and affirmed internationally and regionally find concrete or at least aspirational (legal and practical) application within the domestic realm of the region's states. This will hopefully provide an initial sense of the extent to which an institution developed mostly at the international level is trickling down within states, thus changing their normative landscape and hence behaviour.

Here my analysis takes into consideration the following parameters: (1) the presence of a Ministry of Environment/Ecology/Climate; (2) the inclusion of references to environmental protection in national constitutions; (3) whether there are national documents/programmes addressing climate change and the current environmental crisis; (4) whether references to the environment are included in foreign policy documents; and (5) the climate performance of each state according to data provided by the Environmental Performance Index (EPI).[4] This index uses 32 performance indicators across 11 issue categories, to rank 180 countries on environmental health and ecosystem vitality (highest value 82.5; lowest value 22.6). The EPI 'offers a score card that highlights leaders and laggards in environmental performance and provides practical guidance for countries that aspire to move toward a sustainable future' (EPI, n.d.).

The above reveals, first of all, that all Central Asian states are working towards the creation of an infrastructure of agencies, documents and bodies to be tasked with addressing, not just the political and economic, but also the methodological and epistemological aspects of environmentalism (that is to say, there is a push for the formation of epistemic communities in Central Asia tasked with studying climate change and its impact on the region).

Second, all Central Asian states were already aware of the importance of protecting the environment in the early 1990s, as demonstrated by the insertion of environmental priorities in their constitutions. This is indeed a nice parallel with the international dimension, which showed that, even at the onset of independence, their representatives were addressing international forums to 'sensitise' the international community on green matters and the necessity for financial and technical help. Linked to this, there is the almost uniform presence of 'green principles' in the foreign policy documents of Central Asian states, with the exception of Uzbekistan. This shows that

---

[4] For a detailed outline of the database's methodology, see https://epi.yale.edu/downloads.

'environmentalism' has indeed acquired the status of an institution of international society, as the link between the state as an actor and environmental stewardship as a principle is thereby emphasised in dealings with other members of the international community.

Third, every Central Asian country now has a dedicated body to address environmental issues, although these vary in terms of capacity, budget, and dependency on political power and the broader web of interests within the countries. For example, the minister of Ecology of Kazakhstan, Brekeshev Serikkali Amangaliuly, has previously worked in the oil and gas sector, which should not detract from his commitment to diversification and greening of the economy, but rather shows how it is difficult for these countries to create a new class of environmentally conscious citizens and public servants given the heavily carbon-based background.

Fourth and lastly, while all Central Asian states are placed at the bottom of the EPI ranking, three of them have shown improvements—Kazakhstan, Turkmenistan and Uzbekistan. The attentive reader will notice that these are the three countries of the region that are rich in hydrocarbons, and therefore countries where reforms and green plans are more visible. Yet, they are also the three richest countries in the region. As Weinthal observed twenty years ago, in Central Asia '[Climate change] is no longer just a technical problem; it is now also a political one that ultimately links issues of environmental scarcity and degradation with the political, economic, and social challenges inherent in the transition from communist rule' (2002, 7).

5 Findings, Recommendations and Conclusions

This chapter intended to provide an answer to the question, 'Has environmentalism been institutionalised in Central Asia?' In light of the three-part analysis offered above (international, regional and domestic levels of analysis) the answer is 'yes', although of course, like many things in politics and international relations, this 'yes' hides several tensions, nuances and contradictions.

The first important finding of this chapter is that environmental stewardship in Central Asia is being institutionalised in parallel with a process of adaptation to the Western liberal order and its normative and financial architecture. Given that the conditions post-independence in Central Asia were not ideal for environmentalism, it is understandable that the pace and depth of environmentalism is more pronounced at the discursive level as opposed to on the practical policy level. Yet, what should be seen as a sign of positive compliance with environmental discourses, norms and practices is that the difficulties in adapting the environmental agenda are not a matter of ideology, but rather a matter of capacity. What matters is the lack of funds and infrastructures, as well as a clear plan to devise single payments and compensation for policy shifts to address climate change and environmental crises. In Central Asia, there is an absence of developmentalist narratives and of normative or ideological opposition to environmental stewardship.

The second important finding is that the Central Asian states, when cooperating on environmental matters, seem to act as a region in the international realm, but less so within the region itself. This may be explained by their different strategies and tactics, different needs and different institutional preferences. As the regional analysis showed, this means that the logic of institutionalisation of environmentalism has a strong component of *calculation* as opposed to pure belief. That is, environmentalism needs to be 'talked about' because of its reputation, the need to attract funds, and to ensure minimal compliance with global environmental standards, but without encroaching on other fundamental institutions such as sovereignty or human rights, and with a limited impact on the economy. The consequence of this is that, while there are elements of centralised cooperation on environmental matters in Central Asia (mostly though Chinese actions within the SCO framework) and liberal approaches to climate-change policies, such as inviting international financial institutions and donors, what we observe in Central Asia is a form of middle-ground or 'induced' cooperation based on isolated payments and investments, the role of third parties and regional consultation.

In light of this, by means of a conclusion, two recommendations can be offered to policymakers and stakeholders. The first one is that engagement, dialogue, support and help to Central Asia should continue in technological, scientific, epistemic, infrastructure and financial terms. The analysis above clearly demonstrates that there is awareness, willingness and a need to embrace environmentalism and climate-change related action. The second recommendation is that this support may not necessarily find the best application if framed along exclusively regional lines. While a region-wide approach is important to stress the transnational nature of climate change and environmental degradation, the different challenges, resources, human capital and social contracts in the region lead to differences in complexity of policy design, project feasibility and social priorities. Environmentalism is being institutionalised in Central Asia. The challenge now is for it to move from the sphere of calculation and state financial needs to the realm of belief and moral principle for the benefit of states *and peoples*.

References

Agostinis G, Urdinez F (2021) The nexus between authoritarian and environmental regionalism: an analysis of China's driving role in the Shanghai Cooperation Organization. Probl Post-Communism 69:330–344

Akiner S, Tideman S, Hay J (eds) (2020). Sustainable development in Central Asia. Routledge, London

Akorda (2022) Joint statement following the consultative meeting of the heads of state of Central Asia. https://www.akorda.kz/ru/sovmestnoe-zayavlenie-po-itogam-konsultativnoy-vstrechi-glav-gosudarstv-centralnoy-azii-672511. Accessed 21 Feb 2022

Buzan B (2004) From international to world society? English school theory and the social structure of globalisation. Cambridge University Press, Cambridge

CARECECO (2021) Kyrgyzstan president visits the Central Asian Pavilion at UNFCCC COP26 in Glasgow. CARECECO. https://carececo.org/en/main/news/news/prezident-kr-posetil-pavilon-tsentralnoy-azii-na-ks-26-rkik-oon-v-glazgo-/. Accessed 21 Feb 2022

ECIFAS-TJ (2021) Pavilion of Central Asia within the framework of COP-26. ECIFAS-TJ. https://ecifas-tj.org/en/2021/11/03/pavilion-of-central-asia-within-the-framework-of-cop-26/. Accessed 14 Feb 2022

EEU Treaty (2015) Treaty on the Eurasian Economic Union. https://www.wto.org/english/thewto_e/acc_e/kaz_e/wtacckaz85_leg_1.pdf. Accessed 21 Feb 2022

EPI (n.d.) Environmental Performance Index. https://epi.yale.edu/. Accessed 21 Feb 2022

Falkner R (2021) Environmentalism and global international society. Cambridge University Press, Cambridge

Falkner R, Buzan B (2019) The emergence of environmental stewardship as a primary institution of global international society. Eur J Int Rel 25:131–155

Kamilov A (1994) Speech at the 49th plenary session of the UN General Assembly. A/49/PV.17

Liu W, Liu L, Gao J (2020) Adapting to climate change: gaps and strategies for Central Asia. Mitig Adapt Strat Glob Change 25:1439–1459

Masalieva J (2022) Central Asian countries intend to determine water and energy balance. 24kg. https://24.kg/english/224097_Central_Asian_countries_intend_to_determine_water_and_energy_balance/. Accessed 21 Feb 2022

Mitchell RB (2022) International Environmental Agreements Database Project 2002–2022 (Version 2020.1). http://iea.uoregon.edu/. Accessed 4 Mar 2022

Nazarbayev N (1997) Speech at the 19th special session of the UN General Assembly. A/S-19/PV.1

Osmonalieva B (2021) Sadyr Japarov tells UN General Assembly about unscrupulous investors. 24kg. https://24.kg/english/207915_Sadyr_Japarov_tells_UN_General_Assembly_about_unscrupulous_investors/. Accessed 19 Feb 2022

Paris Agreement (n.d.) The Paris Agreement. https://unfccc.int/process-and-meetings/the-paris-agreement/the-paris-agreement. Accessed 15 Feb 2022

Rahmon E (2021) Address by Mr. Emomali Rahmon, President of the Republic of Tajikistan, at the 76th plenary session of the UN General Assembly. Annex III, A/76/PV.11

Shayakhmetova Z (2021) Kazakhstan and Tajikistan to cooperate as part of Green Bridge Partnership Program to promote climate policies in Central Asian region. The Astana Times. https://astanatimes.com/2021/11/kazakhstan-and-tajikistan-to-cooperate-as-part-of-green-bridge-partnership-program-to-promote-climate-policies-in-central-asian-region/. Accessed 26 Feb 2022

Statement 'Voice of Central Asia' (2021) Правительства Центральной Азии выступили с заявлением на климатическом саммите в Глазго. Avesta - информационное агентство. https://avesta.tj/2021/11/16/pravitelstva-tsentralnoj-azii-vystupili-s-zayavleniem-na-klimaticheskom-sammite-v-glazgo/. Accessed 20 Feb 2022

UNECE (n.d.) EPR reviewed countries. UNECE. https://unece.org/epr-reviewed-countries

Weinthal E (2002) State making and environmental cooperation: linking domestic and international politics in Central Asia. MIT Press, Cambridge, MA

Yu Y, Chen X, Malik I, Wistuba M, Cao Y, Hou D, Ta Z, He J, Zhang L, Yu R, Zhang H, Sun L (2021) Spatiotemporal changes in water, land use, and ecosystem services in Central Asia considering climate changes and human activities. J Arid Land 13:881–890

Vakulchuk R, Daloz AS, Overland I, Sagbakken HF, Standal K (2022) A void in Central Asia research: climate change. Cent Asian Surv: 1–20. https://doi.org/10.1080/02634937.2022.2059447

Zheenbekov S (2020) Address by Mr. Sooronbai Zheenbekov, President of the Kyrgyz Republic, at the 75th plenary session of the UN General Assembly. Annex 11, A/75/PV.6

Filippo Costa Buranelli is Senior Lecturer in International Relations at the University of St Andrews and is currently Chair of the English School section at the International Studies Association. His interests include International Relations theory, international history, global governance, Eurasian politics, and comparative regionalism. His research has been published in Millennium: Journal of International Studies, International Studies Quarterly, International Politics, Geopolitics, International Relations, Central Asian Affairs, and Problems of Post-Communism, among several other outlets. He is currently editing a forum on the politics of informality and global governance for International Studies Review and serves as expert advisor to the Ministry of Foreign Affairs of Italy on the establishment of an Italy–Central Asia consultation mechanism in the field of scientific cooperation

Open Access This chapter is licensed under the terms of the Creative Commons Attribution 4.0 International License (http://creativecommons.org/licenses/by/4.0/), which permits use, sharing, adaptation, distribution and reproduction in any medium or format, as long as you give appropriate credit to the original author(s) and the source, provide a link to the Creative Commons license and indicate if changes were made.

The images or other third party material in this chapter are included in the chapter's Creative Commons license, unless indicated otherwise in a credit line to the material. If material is not included in the chapter's Creative Commons license and your intended use is not permitted by statutory regulation or exceeds the permitted use, you will need to obtain permission directly from the copyright holder.

The Importance of Boosting Societal Resilience in the Fight Against Climate Change in Central Asia

Fabienne Bossuyt

Abstract Drawing on post-development thinking, this chapter argues that resilience strategies involving the sharing of responsibilities among individuals and communities will increase the ability of the Central Asian countries to stand up to the impact of climate change. Given that Central Asian societies have a strong tradition of home-grown solidarity movements and locally embedded practices of self-reliance, governments, as well as major international donors such as the European Union, the World Bank and the UNDP should help boost societal resilience to climate change in Central Asia by supporting the ability of local societal actors to self-organise and draw on their own local strengths and knowledge of available resources and infrastructure.

Keywords Climate resilience · Self-governance · Post-development thinking · Central Asia

1 Introduction

Central Asian countries are among the most vulnerable in the world to the effects of climate change. Climate change has already claimed wide-ranging impacts on livelihoods in Central Asia and is negatively affecting the economy and agricultural outputs (OSCE 2017; Reyer et al. 2017; Sommer et al. 2013). At the same time, the region is recovering from the COVID-19 pandemic, which has had far-reaching socio-economic effects across Central Asia, especially for the most vulnerable parts of the population (Gleason and Baizakova 2020). In this, the COVID-19 crisis has only further revealed the need to prioritise the interconnection between nature, human resilience and sustainable development in any attempt to increase preparedness for further large-scale crises.

The impact of climate change in Central Asia is set to become more pervasive during the decade from 2020 to 2030 (Vakulchuk et al. 2022), with, among other

F. Bossuyt (✉)
Department of Political Science, Ghent University, Ghent, Belgium
e-mail: fabienne.bossuyt@ugent.be

things, more frequent and intense heat extremes, uncertain precipitation patterns, and further glacial melting (OSCE 2017; Reyer et al. 2017). Therefore, states and societies in Central Asia must learn to recover as quickly and efficiently as possible and to handle the impacts of climate change. The concept of resilience is well-placed to capture this, as it is about the extent to which a state or society can mitigate or adapt to crises. Central Asian countries have already taken several important steps to adapt to the harmful effects of climate change. They have all developed national strategies and action plans to fight climate change and move towards low-carbon economies, and have initiated projects on mitigation and adaptation (OSCE 2017). Moreover, the governments of the Central Asian countries have demonstrated readiness and commitment to join forces to address climate change, albeit not always wholeheartedly (OSCE 2017). They are all actively involved in dealing with climate change with direct support from UN agencies, international donors, and financial organisations which are increasingly promoting the notion of resilience as an essential way to be prepared for climate change.

However, as the COVID-19 pandemic revealed, most governments in Central Asia are not prepared to deal with a crisis of such magnitude (Gleason and Baizakova 2020). Moreover, the authorities of the Central Asian countries are not capable of taking (and/or willing to take) the necessary measures to sufficiently adapt to climate change and mitigate its consequences (OSCE 2017; Xenarios et al. 2019). As highlighted in the 2017 Environment and Security Initiative (ENVSEC) report for Central Asia, the financial support offered by international donors may give the Central Asian governments 'the impetus for taking actions, but they also have to develop the potential that is necessary for the effective implementation of these measures. [...]. The countries of the region dispose [only] over basic institutional capacity to plan and implement climate change measures, with some countries having stronger capabilities than others' (OSCE 2017). In addition, as Xenarios et al. (2019) indicate, while the effects of climate change are most severe in mountainous areas, 'the Kyrgyz and Tajik national government structures are characterised by hierarchical decision-making with increasingly authoritarian elements, as well as by inadequate consideration of the specific needs of remote regions like mountain areas in national policy processes'.

Post-development thinking emphasises the value of self-reliance, linked to the notion that people are the agents of their own change and should act based on their local knowledge about what is good for them (see for example Gudynas 2011a, b). Drawing on this school of thought, this chapter argues that resilience strategies involving the sharing of responsibilities among individuals and communities will increase the ability of these countries to stand up to the impacts of climate change. The COVID-19 crisis has put this assertion to the test. In the case of Central Asia, the crisis has revealed that civil society and community-based initiatives were instrumental in addressing the direct impacts of the pandemic (Berdiqulov et al. 2021; Cabar 2021). Similarly, when it comes to standing up to the effects of climate change, there are indications that local civil society actors in Kazakhstan, Kyrgyzstan and Tajikistan are playing an important role by mobilising resources to tackle local climate and environmental measures based on the locally attuned knowledge and skills of local populations (OSCE 2017).

Therefore, given that Central Asian societies have a strong tradition of locally embedded practices of self-reliance, international donors could help boost societal resilience to climate change in Central Asia by supporting the ability of local societal actors to self-organise and draw on their local strengths and knowledge of available resources and infrastructure.

Across Central Asia, major international donors, including the EU, UNDP and the World Bank, have initiated programmes and projects that focus on boosting resilience by increasing adaptation to climate change and mitigation of its consequences (Vakulchuk et al. 2022). However, rather than boosting 'the ability of people or a society to self-organise, drawing on its local strength and knowledge of available resources, and more importantly, on their hope for a better future' (Korosteleva and Petrova 2020, 2), the resilience agenda of most of these international donors reflects a 'liberal internationalist approach of ready-made solutions and a new-liberal working with responsibilised subjects, from a distance' (Korosteleva 2018, 4).

This observation also applies to the way in which the EU promotes climate resilience in Central Asia, and resilience more generally. Since the launch of the EU's Global Strategy in 2016, the EU has come to conceptualise resilience as 'the ability of states and societies to reform, thus withstanding and recovering from internal and external crises' (European Union 2016, 23). In what appears to be a promising feature, the EU acknowledges that strengthening resilience in recipient countries involves granting local societies more ownership over development initiatives, given that 'positive change can only be home-grown' (European Union 2016, 27). Yet, some argue that the EU's understanding of, and approach to, resilience falls short of truly empowering local people and strengthening societal governance from the bottom-up, owing to its continued neoliberal and Eurocentric fixation on EU norms-sharing through ready-made solutions (Joseph and Juncos 2019). This tendency is also manifest in the EU's new Strategy for Central Asia, which prioritises resilience, including enhancing environmental, climate and water resilience (European Commission and HR 2019).

The EU's approach to resilience represents a move away from 'full intervention' and shifts responsibility from the international community to local actors, thereby invoking a progressive discourse of empowerment. However, 'this is done according to a global template that is decided not at the grassroots level, but among international, non-governmental and donor organizations and other international actors' (Joseph and Juncos 2019, 999). While recognising the leading role of partner countries, the EU is 'effectively telling them what their practices should be' (ibid., 1000).

Therefore, this chapter aligns itself with an emergent scholarship that argues that, if international donors are serious about promoting resilience as a way to empower 'the local' and contribute towards a truly sustainable future for Central Asian societies, then these donors will need to embrace a de-centred, post-neoliberal approach to resilience instead of the neoliberal approach that they currently employ (Korosteleva 2018). This implies that donors such as the EU would have to accept 'the other'—namely those Central Asian societies—for what they are, and advocate home-grown self-organisation and self-governance based on a deep understanding of the local meaning of a 'good life' and local knowledge about the available resources (Korosteleva 2018, 2019; Jerez and Morrissey 2020).

As Korosteleva (2018) highlights, donors such as the EU would need to de-centre their approach to societal resilience more radically 'from those who govern to those who are subjectivised by it, and not by way of creating compliant subjects but rather by way of empowering "peoplehoods", seeking to turn their existing capacities into critical infrastructures to necessitate change, from within, and make it sustainable' (ibid., 3). Indeed, these international donors uphold an understanding of resilience as neoliberal governance, which boils down to building resilience 'outside-in', namely by providing external solutions to local problems for societal groups and communities in recipient countries who are made into compliant subjects and consumers of European/Western practices of good governance (Korosteleva 2018).

To further advance this nascent body of literature, this chapter brings in insights from post-development thinking as a way to conceptualise and operationalise a de-centred, post-neoliberal paradigm of societal resilience to climate change.

2 Post-development Thinking

Post-development thinking has been key in challenging the underlying assumptions and implications of classical Western development theory, as well as the neoliberal paradigm that has determined international development assistance. In providing feasible alternatives to the neoliberal paradigm, post-development thinking has, *inter alia*, proposed the notion of 'good life'. The term 'good life' comes from the Latin American concept *Buen Vivir*, which can be translated as 'good living' or 'living well'. The concept arose in Latin America in response to the neoliberal development strategies that governments and multilateral development banks were following (Gudynas 2011b). By putting into question the reductionism of classical Western development theory, which reduces development to economic growth, *Buen Vivir* foregrounded quality of life that goes beyond consumption or property, by connecting it to the collective well-being of humans and by considering well-being to be possible only within a community (Gudynas 2011a, b). The concept brings together a set of ideas that act not only as a critique of Western development thinking—with its ideology of progress and emphasis on economic growth—but also as an alternative to those conventional notions of development (Gudynas 2011b). It is in this regard that *Buen Vivir* has been connected to post-development thinking, a collection of ideas invigorated by Arturo Escobar in the 1990s (e.g. Escobar 1995), and which has called for abandoning capitalism and the Western-centric development paradigm by advocating locally inspired alternatives to development (Schöneberg 2016).

By offering alternatives to development that emerge from indigenous traditions, *Buen Vivir* provides possibilities to move beyond the Eurocentric tradition (Gudynas 2011a). Related concepts can be found in other parts of the Global South. These local meanings of 'good life', despite coming from different parts of the Global South, have in common an emphasis on the value of self-reliance, linked to the notion that people are the agents of their own change and should act based on their local knowledge about what is good for them. In other words, these local understandings of the good life and well-being are strongly reflected in local forms and practices of self-reliance and self-governance, which are key to ensuring resilience at the community level.

Although local meanings of 'good life' in Central Asia have not yet been studied widely in the academic literature, it is possible to draw some of the key features of a 'good life' in the region based on an existing scholarly knowledge of the region. In Central Asia, and especially in rural and mountainous areas, a good life is closely associated with the moral principle of trust, as reflected in 'trust networks', which secure community members' good life through reciprocal practices (Boboyorov 2013). These support networks are conditioned by reciprocity, both as a moral good and as material aid in emergencies (ibid.). They run within families, neighbourhoods (*mahallas*) and villages, and across lines of kin, and they 'involve traditional forms of community interaction, management and positions of responsibility' (Earle 2005).

Similar to understandings of good life and well-being that can be found in other parts of the Global South, the centrality of social trust and solidarity as the basis for a good life and well-being are strongly reflected in local forms and practices of self-reliance and self-governance in Central Asia. In Kyrgyzstan and Uzbekistan, for instance, at the community level a central role is taken up by the *aksakals*, or elders, typically well-respected older male members of the community who are given a leading role in the community (Earle 2005; Kreikemeyer 2020). *Aksakals* are seen as symbolising a caring civic community that offers alternative values to the neoliberal values espoused by markets and elites (Satybaldieva 2018). At the community level, there is also the traditional practice of *khashar*, also known as *ashar*, a form of collective voluntary work, in which people from the community are expected to provide assistance for community members as part of a joint effort to improve living standards within the community (ibid.).

The extent to which local forms and practices of self-organisation and self-reliance in Central Asia act as functions of resilience-building was vividly illustrated during the COVID-19 pandemic (Gleason and Baizakova 2020). As similar crises are likely to emerge in the future, societal resilience will need to be significantly enhanced in order to cope with the shocks that such crises create. This applies not least to the case of climate change, as its devastating effects are only set to increase. I argue that resilience strategies involving the sharing of responsibilities among individuals and communities will increase the ability of Central Asian countries to respond to major crises such as climate change.

In Central Asia, especially in Kazakhstan, Kyrgyzstan and Tajikistan, reports suggest that local civil society actors have been playing a conducive role in fighting climate change by mobilising resources to take local climate and environmental measures thanks to their awareness of the locally attuned knowledge and the capacities of local people (OSCE 2017). However, to date, we lack further concrete information on this as no research has yet been conducted that systemically explores the role that local communities and organisations play in addressing climate change and environmental challenges in Central Asia. Therefore, in the remaining part of this chapter, a brief exploratory analysis will be offered on how home-grown forms of self-governance and self-organisation that embody an indigenous understanding of the good life are building climate resilience in Central Asia.

3 Local Self-Governance to Boost Climate Resilience in Central Asia

The mountainous areas in Central Asia have proven particularly vulnerable to the effects of climate change (Hughes 2012; Xenarios et al. 2019). One of the main issues is the degradation of pastures as a result of changes in precipitation, which is often further exacerbated by excessive grazing (Wang et al. 2020). Importantly, local initiatives at the community level are proving pivotal in responding to these climate pressures and fostering grassland restoration. In Kyrgyzstan, a recent study found that, while almost 60% of the country's pasturelands continue to experience grassland degradation, 40% of Kyrgyzstan's grasslands have actually expanded, as a direct result of community-level activities focused on grassland recovery (Wang et al. 2020).

Across the mountain regions in Kyrgyzstan and Tajikistan, several cases can be found of local forms of self-governance and self-organisation that seek to reduce the environmental vulnerability of the local communities and adapt pastoralism to the effects of climate change. In Kyrgyzstan, such initiatives started to emerge from the early 2000s onwards and have become more prominent since the enactment of the Kyrgyz Republic Law of Pasture in 2009. By introducing this law to enhance the sustainability of pasture use, the Kyrgyz government sought to overcome the problem of fragmentation and rehabilitate rangeland by decentralising the policy and delegating responsibility for managing pastures to community-based user units (Wang et al. 2020). This decentralisation process included the establishment of democratically elected pasture committees, which are mandated to oversee the management of the pastures (Hughes 2012). Since then, the region has witnessed the emergence of various types of local organisations which are involved in the sustainable management of grasslands, including community-based organisations, such as the so-called 'Village Organisations' and other types of civil society organisation (CSO). As Hughes (2012, 108) reported, this prompted 'a heightened sense of community and sense of responsibility for the stewardship and ownership of resources', which 'is apparent when herders and community members sit together to coordinate their moves to summer pastures, repair bridges by mobilizing their own resources or develop and implement pasture use plans'.

Among the CSOs that emerged in the wake of these developments in Kyrgyzstan are mountain-focused non-governmental organisations (NGOs), such as the Institute for Sustainable Development Strategy (ISDS) Public Fund, as well as various mountain associations. The NGOs tend to act as bridges between various levels of stakeholders, connecting national governments to village institutions and the general public. According to Hughes (2012, 93), 'they often communicate "mountain voices", advocate for interactive and open processes of policy formulation and act to bridge any gaps between new legislation and strategies and the realities in mountain communities'. Mountain associations, in turn, focus on the promotion of mountain environmental knowledge and the cultivation of responsible outdoor traditions, and they organise activities such as picking up litter (Hughes 2012).

At the community level, there is also the Alliance of Central Asian Mountain Communities (AGOCA). This association, which was created in 2003, is based in Kyrgyzstan and also includes mountain villages in Kazakhstan and Tajikistan. The alliance consists of citizen associations at the community level that are working towards improving the living conditions of mountain communities, including their capacity to adapt to the pressures of climate change (Kohler and Maselli 2012). They act as self-governance bodies that implement development projects and 'communicate needs, ideas, and visions to state representatives at the local level, and negotiate with them' (ibid., 41).

Together, this wide range of local organisations and forms of self-governance at the community level are playing a crucial role in enhancing societal resilience to climate change in the mountain areas of Central Asia. Importantly, the emergence of this locally embedded involvement in addressing the effects of climate change is being increasingly accompanied by a revival of traditional knowledge and practices. In the specific case of pasture management, there is a tendency to advocate the revival and preservation of traditional pastoralism practices to reduce the vulnerability of ecosystems as a result of climate change. For instance, a recent project overseen by the ISDS Public Fund together with the Cholpon Pasture Users Union specifically promoted the integration of traditional nomadic knowledge and practices into community-based pasture conservation in the Cholpon rural municipality (ISDS 2019). Situated in the north of Kyrgyzstan, this area has been harshly affected by the combination of excessive grazing and irregular precipitation due to climate change. Having abandoned traditional pastoral knowledge and practices for decades, local pastoralists are once again increasingly relying on the traditional knowledge and practices inherited from their ancestors in order to sustainably maintain their livelihoods and enhance their resilience to the effects of climate change (ISDS 2019).

Importantly, some of these initiatives have been successfully supported by international donors. Indeed, the creation of AGOCA was an outcome of the Global Mountain Summit held in Bishkek in 2002. The summit was the closing event of the UN's International Year of Mountains, which sought to promote 'the conservation and sustainable development of mountain regions, thereby ensuring the well-being of mountain and lowland communities' (Nikonova et al. 2007, 24). However, as stakeholders acknowledged that the participatory process provided hardly any space for the voices of the mountain communities themselves, the Central Asian Mountain Partnership (CAMP), a programme created in 2000 and funded by the Swiss Agency for Development and Cooperation, decided to organise the first Conference of Mountain Communities for Sustainable Development as a pre-summit event (Nikonova et al. 2007). As a major outcome of the second conference that took place in Tajikistan the following year, AGOCA was created as a way to stimulate mountain communities to organise themselves in order to stand up to the environmental pressures and effects of climate change (ibid.).

CAMP itself was created by the Swiss Agency for Development and Cooperation to promote sustainable mountain development in Central Asia 'by encouraging a more economically, ecologically and socially sustainable use of resources, through different stakeholders in Kyrgyzstan, Tajikistan, and Kazakhstan' (SDC 2009). Together with other international development players, CAMP has stimulated local communities to take matters into their own hands in order to enhance their resilience to environmental challenges and climate change. Among other things, by relying on participatory and community-based approaches, CAMP has contributed to improved pasture management practices in Kyrgyzstan by introducing a flexible pasture management system (Hughes 2012).

Importantly, both through AGOCA and through projects funded by CAMP and other international development players, a specific role in these endeavours is played by locally embedded organisations and home-grown forms of self-governance that reflect an indigenous understanding of the 'good life' as conceptualised in the preceding section. Indeed, as reported by Nikonova et al. (2007), in the case of AGOCA, the traditional local governance bodies *ayul okmoty* in Kyrgyzstan, *jamoat* and *hukumat* in Tajikistan, and *akimat* in Kazakhstan fulfil an important function in enhancing community resilience thanks to their capacity to mobilise the local community, including in the form of the traditional custom of *khashar* (*hashar* in Tajikistan, *assar* in Kazakhstan), which is still being practised in mountain regions in Central Asia as a form of collective action (see above).

In the majority of environmental projects involving local partners, these traditional governance bodies have played a crucial role. This was the case, for instance, in the transboundary Pamir-Alai Land Management (PALM) project, which was funded by the Global Environment Facility (Hughes 2012), and in the CAMP Kuhiston project in Tajikistan (ibid.). The latter project, which sought to link disaster risk management with the planting of suitable fruit tree species to enhance land productivity in Nurobod District in central Tajikistan, benefited from the involvement of the *jamoat* (ibid.). Among other things, the head of the village initiated a *khashar* to mobilise the local population to erect a wire fence and to plant saplings (ibid.).

4 Conclusion

Drawing on post-development thinking, which emphasises the value of self-reliance linked to the notion that people are the agents of their own change and should act based on their local knowledge about what is good for them, this chapter has argued that resilience strategies involving the sharing of responsibilities among individuals and communities will increase the ability of the Central Asian countries to stand up to the impacts of climate change.

The extent to which local forms and practices of self-organisation and self-reliance in Central Asia help to build resilience has been vividly illustrated during the COVID-19 pandemic. Indeed, during the pandemic, civil society and community-based initiatives across Central Asia were instrumental in addressing the direct impacts of the pandemic, especially in areas where governments fell short, such as medical support and the provision of information and social protection. While the COVID-19 pandemic has revealed and exacerbated existing challenges in Central Asia relating to, among other things, poor state governance and weak state capacities, it has highlighted the key role that grassroots civil society and community-based practices of self-reliance play in strengthening resilience in the face of a major crisis.

Much more than in Western societies, societies in Central Asia are collective in nature rather than individualistic. This also implies that solidarity among members of the community is much more embedded in local practices and customs than in Western societies. Moreover, in Central Asia, state capacity is not always strong enough to cope with a crisis of great magnitude, and vital public services are deficient. As we have seen above, this also applies to the governments' capabilities of addressing the effects of climate change. As similar crises are likely to emerge in the future, societal resilience will need to be significantly enhanced in order to cope with the shocks that such crises create. This applies not least to the case of climate change, as its devastating effects are only set to increase.

In Central Asia, major international donors like the EU, the World Bank and the UNDP should therefore draw lessons from the societal responses in the region to the COVID-19 crisis, and thus support more actively the ability of local societal actors to self-organise and draw on their own local strength and knowledge of available resources. Our brief exploratory analysis has shown how home-grown forms of self-governance and self-organisation in the mountainous areas of Central Asia are already building resilience to climate change. Some international donors, such as the Swiss Agency for Development and Cooperation, have already adopted more locally owned and locally driven approaches to enhancing climate resilience in Central Asia. Their approach appears to endorse the post-development notion that people are the agents of their own change and should act based on their local knowledge and capacities. The other major donors in the region that are promoting climate resilience should follow suit. Indeed, instead of moulding resilience-building externally, these international donors need to acknowledge that resilience-building starts internally, from the communities, drawing on their existing resources and knowledge and their local understanding of 'good life'.

References

Berdiqulov A, Buriev M, Marinin S (2021) Civil society and the COVID-19 governance crisis in Kyrgyzstan and Tajikistan. Institut für Europäische Politik, Berlin

Boboyorov H (2013) The ontological sources of political stability and economy: Mahalla mediation in the rural communities of Southern Tajikistan. Crossroads Asia Working Paper Series, 13, Bonn

Cabar (2021) State and civil society during the COVID-19 pandemic in Central Asia, September 27. https://cabar.asia/en/state-and-civil-society-during-the-covid-19-pandemic-in-central-asia.

Earle L (2005) Community development, 'tradition' and the civil society strengthening agenda in Central Asia. Cent Asian Surv 24(3):245–260

Escobar A (1995) Encountering development. The making and unmaking of the third world. Princeton University Press, Princeton

European Commission and High Representative of the Union for Foreign Affairs and Security Policy (2019) Joint Communication on the EU and Central Asia: new opportunities for a stronger partnership, Brussels, May15.

European Union (2016) Global strategy for the European Union's foreign and security policy: 'shared vision, common action: a stronger Europe', Brussels, June

Gleason G, Baizakova K (2020) COVID-19 in the Central Asian region: national responses and regional implications. Connect: Q J 19(2):101–114

Gudynas E (2011a) Buen Vivir: today's tomorrow. Development 54:441–447

Gudynas E (2011b) Good life: germinating alternatives to development. America Latina en Moviemento, issue 462

Hughes G (ed) (2012) University of Central Asia, Zoï Environment Network, Mountain Partnership, GRID-Arendal. 2012. Sustainable Mountain Development. From Rio 1992 to 2012 and beyond. Central Asia Mountains. https://ucentralasia.org/media/bllawjt1/web-caf-central-asia-mountains.pdf

ISDS—Institute for Sustainable Development Strategy Public Fund (2019) Traditional knowledge for climate resilience: collaborative strategies to mitigate vulnerability and enhance adaptation of pastoralism to climate change. https://satoyama-initiative.org/wp-content/uploads/2019/06/ISDS_IPSI-Case-Study-Summary-Sheet-web-min.pdf

Jerez Y, Morrissey J (2020) Subaltern learnings: climate resilience and human security in the Caribbean. Territ Polit Gov. https://doi.org/10.1080/21622671.2020.1837662

Joseph J, Juncos AE (2019) Resilience as an Emergent European Project? The EU's place in the resilience turn. J Common Mark Stud 57(5):995–1011

Kohler T, Maselli D (eds) (2012) Mountains and climate change—from understanding to action. Bern, Geographica Bernensia

Korosteleva E (2018) Paradigmatic or critical? Resilience as a new turn in EU governance for the neighbourhood. J Int Relat Dev. https://doi.org/10.1057/s41268-018-0155-z

Korosteleva E (2019) Reclaiming resilience back: a local turn in EU external governance. Contemp Secur Policy. https://doi.org/10.1080/13523260.2019.1685316

Korosteleva E, Petrova I (2020) Resilience is dead. Long live resilience? Italian International Affairs Institute. https://www.iai.it/en/pubblicazioni/resilience-dead-long-live-resilience?fbclid=IwAR3PTlugF1wKU0qJtkzaoJFxziDqVpEMdOU7roRAin3dD9uFJEU6HL9y1WQ.

Kreikemeyer A (2020) Local ordering and peacebuilding in Kyrgyzstan: what can customary orders achieve? J Interv Statebuilding. https://doi.org/10.1080/17502977.2020.1758425

Nikonova V, Rudaz G, Debarbieux B (2007) Mountain communities in Central Asia: networks and new forms of governance. Mt Res Dev 27(1):24–27

OSCE (2017) Climate change and security in Central Asia. Regional Assessment Report. https://www.osce.org/secretariat/355471

Reyer C, Otto IM, Adams S et al (2017) Climate change impacts in Central Asia and their implications for development. Reg Environ Chang 17:1639–1650

Satybaldieva E (2018) Working class subjectivities and neoliberalisation in Kyrgyzstan: developing alternative moral selves. Int J Polit Cult Soc 31:31–47

Schöneberg J (2016) Making development political. NGOs as agents for alternatives to development. Baden-Baden, Nomos

SDC—Swiss Agency for Development and Cooperation (2009) Central Asia Mountain Partnership Programme (CAMP). https://www.aramis.admin.ch/Texte/?ProjectID=23960.

Sommer R, Glazirina M, Yuldashev T et al (2013) Impact of climate change on wheat productivity in Central Asia. Agric Ecosyst Environ 178:78–99

Vakulchuk R, Daloz AS, Overland I, Sagbakken HF, Standal K (2022) A void in Central Asia research: climate change. Cent Asian Surv: 1–20. https://doi.org/10.1080/02634937.2022.2059447

Wang Y, Yue H, Peng Q, He C, Hong S, Bryan BA (2020) Recent responses of grassland net primary productivity to climatic and anthropogenic factors in Kyrgyzstan. Land Degrad Dev 31:2490–2506

Xenarios S, Gafurov A, Schmidt-Vogt D et al (2019) Climate change and adaptation of mountain societies in Central Asia: uncertainties, knowledge gaps, and data constraints. Reg Environ Chang 19:1339–1352

Fabienne Bossuyt is Associate Professor and coordinator of the Ghent Institute for International and European Studies at the Department of Political Science at Ghent University (Belgium). She is also co-director of the Eureast Platform of Ghent University. In addition, she is a professorial fellow at UNU-CRIS and an associate researcher at EUCAM. Her main area of expertise is the EU's relations with Central Asia. Her most recent research projects focus on various aspects of the EU's relations with and policies towards Central Asia and other post-Soviet countries, including development policy, human rights promotion and connectivity. She has published articles in, among others, Democratization, Cambridge Review of International Affairs, Journal of International Relations and Development, Eurasian Geography and Economics, Southeast European and Black Sea Studies, East European Politics & Societies, and Communist and Post-Communist Studies. She recently co-edited two books, namely "Principled Pragmatism in Practice: The EU's Policy towards Russia after Crimea" (Brill) and "The European Union, China and Central Asia. Global and Regional Cooperation in A New Era" (Routledge). In the past few years, she has also acted as rapporteur for the EU Special Representative to Central Asia.

Open Access This chapter is licensed under the terms of the Creative Commons Attribution 4.0 International License (http://creativecommons.org/licenses/by/4.0/), which permits use, sharing, adaptation, distribution and reproduction in any medium or format, as long as you give appropriate credit to the original author(s) and the source, provide a link to the Creative Commons license and indicate if changes were made.

The images or other third party material in this chapter are included in the chapter's Creative Commons license, unless indicated otherwise in a credit line to the material. If material is not included in the chapter's Creative Commons license and your intended use is not permitted by statutory regulation or exceeds the permitted use, you will need to obtain permission directly from the copyright holder.

The Culture of Recycling, Re-use and Reduction: Eco-Activism and Entrepreneurship in Central Asia

Aliya Tskhay

Abstract While national governments are setting net-zero targets and drawing up strategies to reduce CO_2 emissions and meet their Paris Agreement commitments, local entrepreneurs are putting words into action. This paper reviews the waste management situations and national strategies across Central Asian countries in addition to the various enterprises which have sought to improve their local environment and change people's behaviour and attitudes towards the environment through waste-management initiatives. The case studies include a zero-waste grocery store; the collection, recycling and re-purposing of waste; and the organisation of events to encourage the reduction of consumption and associated waste. These enterprises combine their business models with social awareness programmes to promote more sustainable lifestyles and environmentally friendly habits. The micro-level focus provides an opportunity to see how entrepreneurs and activists in Central Asian cities are filling the gap left by national governments in promoting sustainability.

Keywords Eco-activism · Eco-entrepreneurship · Central Asia · Waste management · Climate change

1 Introduction

Good waste management is integral to environmental protection and climate change mitigation. In Central Asia, this issue has not received much attention or priority (Vakulchuk et al. 2022). For instance, Kyrgyzstan does not have an institutional structure for waste processing and recycling, and this complicates political change and the implementation of relevant policies (Vorotnikov 2018). Moreover, with a growing population in the region, especially in urban centres, the amount of waste is also growing. Thus, waste management is set to become an even more acute environmental problem in Central Asian countries. The waste accumulated in landfills

---

A. Tskhay (✉)
University of St Andrews, St Andrews, Scotland, UK
e-mail: at99@st-andrews.ac.uk

produces the greenhouse gases CO_2 and methane which are responsible for air pollution and climate change. Air pollution from landfill incineration and water contamination from landfills also result in increased dangers to public health, sanitation and wellbeing (United Nations Environment Programme 2017). It is therefore crucial to address the topic of waste management in this edited volume on climate change and Central Asia.

This chapter looks at the issue of household solid waste management—an area where Central Asian governments have progressed quite slowly in implementing change. At the same time, a wide variety of local initiatives by eco-activists, eco-entrepreneurs and environmentally conscious businesses have spread and provide services for waste recycling and utilisation, while also promoting a culture of environmental awareness in the region. These initiatives are the focus of this paper and can be categorised as recycling, processing and/or social awareness initiatives.

The chapter provides an overview of national policies and strategies for waste management in Central Asian countries and discusses related issues. It then presents case studies of local initiatives led by eco-activists and eco-entrepreneurs to fill gaps in government provisions based on public reports and online information sources such as the websites of the companies, open-access reports or documentary films. The material gathered for this chapter is from secondary sources, and the case examples were picked to conform with the focus on waste management, education and zero waste initiatives.

The chapter seeks to demonstrate that waste collection, recycling and re-use, and associated behavioural change, are possible through bottom-up initiatives, despite the considerable gap in government action on waste management at the national level across the region. As governments struggle with funding and institutional bases for the promotion of waste management, local initiatives pave the way to bring innovative solutions to waste issues, especially in urban areas.

2 Central Asian States and Waste Management

Waste management is not a new concept in the Central Asian states. Indeed, during the Soviet era, waste collection and recycling were encouraged and incentivised. The collection of metal, paper, glass and even food waste was widespread, with centralised strategies and infrastructure (United Nations Environment Programme 2017). After the collapse of the Soviet Union, this fell apart, collection and recycling centres closed down, and the culture of waste utilisation was lost.

Current household waste management systems usually involve the collection of waste in non-segregated bins, with disposal at landfills on the outskirts of municipal centres.[1] The growing urbanisation and population growth in Central Asian countries have put further strain on waste management systems. For instance, in Bishkek alone,

---

[1] This is related to household waste only. The management of industrial waste is administered through a different system, and is beyond the scope of this chapter.

1,000 tonnes of waste are collected daily, and this figure is growing by 20% annually (United Nations Development Programme 2021). In 2018, Uzbekistan recycled only 20% of its solid waste, with the remainder going into landfills and total waste in that year amounting to 80 million tonnes (Eurasianet 2021).

There are two categories of landfill in Central Asia: legal (publicly organised) and illegal (sporadic and unorganised). It is important to underline such a distinction as it also indicates part of the problem in waste management. Recycling is mostly done by private companies and the collection of recyclable waste is limited across Central Asia (United Nations Environment Programme 2017). Several issues need to be highlighted to better illustrate the current situation.

First, the limited national policies and strategies, or lack thereof, present the first challenge to sustainable waste management. The approach to waste management at the national policy level differs between the Central Asian states, which have varied capacities and legislative bases for waste management. In Kyrgyzstan, waste management (both household and industrial) is featured in some legislative acts but without concrete targets, goals or quotas for waste recycling, which has brought criticism from some activists that the government has not seriously considered this issue (Eurasianet 2021). In Kazakhstan, waste management has been acknowledged and is covered in the Environmental Code, which came into force on 2 January 2021 (Ministry of Ecology 2021). The Law on industrial and consumption waste from 2005 regulates Tajikistan's waste management (Majilisi Oli 2011). Uzbekistan has a range of normative acts and laws that regulate the management of industrial and household waste (President of the Republic of Uzbekistan 2020). Turkmenistan is currently developing a national strategy for waste management together with international donors (EBRD and UNDP) (United National Development Programme 2021a).

Second, official data on the numbers of landfills, waste volumes and recycling rates is very limited or non-existent (Wood 2019). This prevents more comprehensive waste management planning, but more importantly, obscures the picture of the range of challenges faced by governments and citizens. In addition, the lack of information and awareness thwarts efforts to educate the public on waste problems. The absence of any baseline data also challenges the monitoring of recycling rates.

Third, the organisation of waste collection is facing numerous issues in Central Asian states. With the lack of segregated bins for different types of waste, household waste goes to common bins, which are then emptied into landfills. The lack of funding for waste collection is an issue, as it is operated at the municipal level. For instance, the Bishkek authorities lack the capacity to organise separate waste collection (Wood 2019). This is also coupled with the social issues arising from unregulated waste collection. Thus, for example, in Kyrgyzstan hundreds of people gain their livelihood from regularly scavenging through urban bins and landfills for plastic, metals and recyclable materials (Rickleton 2010; Eurasianet 2021). The health hazards of this occupation are highly concerning. In addition, criminal groups also organise a waste collection to profit from the black market for such materials as metals and electronics.

Fourth, Central Asian states have limited capacity, knowledge and funding to modernise their waste management systems, and thus rely on cooperation and support from international donors. Regional states cooperate with a range of partners, from UN institutions to international financial institutions—including the World Bank, the Asian Development Bank (ADB) and the European Bank for Reconstruction and Development (EBRD)—and various civil society organisations. Such cooperation is especially targeted to improve policies, bring in knowledge, support infrastructure and provide funding (United Nations Environment Programme 2017). International financial institutions are working on waste management in all five Central Asian republics, which demonstrates that waste management is a problem shared by the entire region.

The above-listed issues demonstrate the profound nature of the waste management problem in the region. The limitations of government approaches to the waste problem are indicative of the amount of work that is required. In other words, there is a critical gap in waste management efforts.

3 National Policies on Waste Management in Central Asia

In general, the approach to waste management in Central Asia is best described as 'out of sight, out of mind'. Waste is dumped on land close to the cities, with some landfills operating beyond their lifespan. The United Nations Environment Programme (UNEP) also acknowledges the poor organisation of landfills, 'with inadequate planning and engineering, no waste sorting or inventories and lacking in modern measures to make them safer' (2017, 16).

National policies on waste management are, however, evolving around the three principles of recycling, reduction and re-use. This requires sufficient infrastructure to support the collection of segregated waste and its further recycling, as well as using recycled materials as part of a circular economy while educating the population to reduce waste through various behavioural changes. These actions are in line with international waste management practices.

In 2013, the President of Kazakhstan signed the 'Concept on the Transition to a Green Economy', a policy document that sets out the vision and direction for the transformation of the economy towards environmentally conscious solutions and programmes (Decree of the President of the Republic of Kazakhstan 2013). The division of waste into categories for the collection was legally stipulated in 2016, and the new Ecological Code of 2021 further defines the categories of waste into wet (food, organic and other waste) and dry (paper and carton, glass, plastic), with bins labelled accordingly. Yet, although the legislation in Kazakhstan matches international standards on waste management, the infrastructure to support it is poor or obsolete (United Nations Environment Programme 2017). Local municipalities are slow to introduce the required waste bins and provide the necessary training to local populations in major cities (2021).

In Kyrgyzstan, the National Development Strategy for 2040 has a provision on waste management and the minimisation of environmental impacts from local businesses (Independent Eco-Expertise). Thus, waste management has been picked up by the government's 'green' agenda. The country receives significant support from international financial institutions and donors for the improvement of solid waste management. The EBRD has been operating a range of programmes since 2013, through loans to support education, infrastructure and development of a legislative base for waste recycling in Bishkek and Osh (European Bank for Reconstruction and Development 2013; Usov 2020). Notably, the EBRD set out to establish the first landfill in Bishkek that would comply with EU standards. However, corruption and misallocation of resources have been reported in connection with this project (Barvinskaya 2019; Wood 2019).

Tajikistan has a similar situation, with a lack of infrastructure, funding and capacity for waste collection and recycling. The EU and the EBRD are providing assistance to the government to facilitate recycling and improve landfill infrastructure (Kukula 2018). The country is also experiencing a problem with the utilisation of chemical waste, such as from lithium batteries (Karimov 2021). Together with UN institutions, Tajikistan's government is taking steps to improve the situation, and in doing so to comply with international conventions, such as the Rotterdam Convention relating to international trade in hazardous chemicals and the Minamata Convention on Mercury (United Nations Environment Programme 2019).

Turkmenistan is to improve its waste management through a joint programme with UNDP called 'Sustainable Cities in Turkmenistan: Integrated Green Urban Development in Ashgabat and Awaza' (United National Development Programme 2021). As part of this programme, the country is working with its partners to develop a national waste management strategy. The goals of the programme are to expand recycling and reduce waste (UNDP in Turkmenistan 2016).

Urban solutions to the waste management problem are also acknowledged as necessary in Uzbekistan. In 2020, the law on solid and construction waste management in Tashkent was introduced, providing a gradual introduction of household waste bins and the segregation of recyclable and non-recyclable waste (UZDaily 2020). This follows the implementation of the Strategy for solid waste management during the 2018–2019 period (President of the Republic of Uzbekistan 2020). This legislation notably provides the foundation for private businesses to operate recycling services in the country, which will be discussed in the next section.

In sum, the Central Asian states are facing very similar issues in addressing waste management. Limited financial capacities, obsolete and poor infrastructure, and a lack of legislation, capacity and knowledge all challenge the functioning of sustainable waste management systems and, thus, endanger public health and the environment across the region. It is not surprising that the international donor community is currently playing a prominent role in filling these gaps. This has been the case since the 1990s when environmental cooperation between donors and local NGOs also provided capacity building programmes on waste management (Weinthal and Watters 2010).

4 Eco-Activists and Eco-Entrepreneurs in Central Asia

Although national policies and strategies in Central Asia are targeted at improving waste management, they have only been introduced in the past ten years and lack swift implementation. Support from the international donor community has helped to reduce the bureaucratic and financial limitations that governments face. At the same time, civil society and entrepreneurs have been actively promoting the culture of reducing, recycling and re-using household waste and facilitating waste management. This section will describe some of the initiatives in Central Asia that are dedicated to the three dimensions (reduce, recycle and re-use), while also educating communities on waste sorting. It is helpful to define who is included under eco-activists and eco-entrepreneurs. Eco-activists are people who campaign to promote sustainable lifestyles and support non-profit organisations dedicated to environmental causes. Eco-entrepreneurs are people who are involved in commercial projects with an environmental agenda, such as recycling businesses. Eco-entrepreneurs can also be eco-activists, especially if they participate in social projects dedicated to environmental issues.

This chapter uses the initiatives below as example cases. What is important to illuminate is the space that these initiatives occupy and the role they are playing in promoting sustainable and environmentally friendly lifestyles. The initiatives cover three key areas—educational projects, social (not-for-profit) projects, and commercial projects—and there is some overlap between these areas.

In order to have a well-functioning waste management system, especially for household waste, educational programmes are particularly important in explaining the correct way to recycle materials. In this respect, eco-activists in Central Asia have developed various projects to explain and assist the general public with waste recycling. An interesting educational tool is the Tazar app for mobile phones in Kyrgyzstan, which was developed by eco-activist Aynura Sagyn. The app provides a map showing the nearest recycling centres in Bishkek, Osh, Karakol, Naryn and Talas. The user can also book waste collection from their home via the app. The app also explains various recycling labels, thus serving as an educational tool for the users to learn more about various types of waste and their proper recycling methods. Sagyn first launched the application for plastic waste collection, but it currently also includes information on glass, paper and clothes recycling (Eurasianet 2021). At its core, the Tazar app facilitates more sustainable lifestyles by connecting consumers with waste recyclers.

In Tajikistan, a student-led initiative called Green Community, founded by eco-activist Anisa Abibulloeva, has partnered with USAID to publish a series of books with short stories on the environment and sustainable living (University of Central Asia 2019). The books will be used in local elementary schools to teach children about climate change and environmental protection. This is a significant step, as climate change and environmentalism are not officially part of the curriculum.

Eco-entrepreneurship is on the rise in Central Asia, and with countries' green agendas and commitments to carbon emissions reduction, waste recycling is becoming a viable business model. The development of relevant legislation and investment from international donors have also stimulated the development of a recycling culture. Yet, the ventures of Central Asia's eco-entrepreneurs are far more progressive than government recycling programmes and were launched long before waste management was on government agendas. For example, 'Fluid' a company that processes organic waste into biogas and fertiliser has been operating in Bishkek since 2002 (2022). The company is not only the first organic waste collector in Kyrgyzstan, but also the first one to utilise waste for recycling and conversion into biogas, which is then used for heating purposes. This venture is reducing emissions locally and, on a national scale, can contribute to Kyrgyzstan's Paris Agreement. 'Fluid' also participates in knowledge and technology sharing with other countries in Eurasia and advises the government of Kyrgyzstan and international organisations. The company's founder admits that proper separation of organic waste by local businesses (such as restaurants) remains a challenge (Eurasianet 2021).

Waste management and concerns about the environment unite different businesses and bring them together to find solutions. Kazakhstan Waste Recycling (KWR) is the largest recycling company in Kazakhstan. Aside from its recycling business of glass, plastic, paper and aluminium, KWR partners with other leading companies to facilitate a recycling culture. It partners with the grocery delivery service Arbuz.kz, construction materials and furniture seller Leroy Merlin, and a restaurant chain Parmigiano Group for the collection of recyclable waste (KWR 2022). KWR is also participating in educational projects to explain to the local population about different types of waste and how to recycle these through its social media channels and its recycling collection points. Similar partnerships can be observed as part of the Hasharweek project in Uzbekistan which engages with local businesses, such as vendors and restaurants, to popularise and encourage recycling (Hashar 2020).

Initiatives to reduce waste are also being developed in Central Asia. Shops such as HelloEco in Kazakhstan and LaLavande in Uzbekistan are demonstrating how it is possible to sell household products with zero-waste packaging, meaning that customers bring their own refillable containers to shop in these stores (Ybyshova 2022).

5 Conclusion

Waste management is a serious issue in Central Asia and poses multiple threats to public health and the environment. The growing population of the region is generating more waste that in general accumulates in landfills. The poor management of landfills creates a string of environmental issues, such as land, water and air pollution, and an increase in methane and CO_2 emissions.

The Central Asian states recognise the problem that waste management poses for the future of the wellbeing of their populations. Multilateral and bilateral projects and programmes have been set up with international financial institutions and donors to tackle the issue and improve waste management. These programmes set out to fill the gap and build the capacity of local governments to operate sustainable waste management systems. All Central Asian states are pursuing national strategies to improve waste management. However, they are experiencing various degrees of success and depths of implementation.

Waste management is currently dominated by private initiatives, led by eco-activists and/or eco-entrepreneurs. The initiatives cover such areas as waste recycling, waste collection and re-purposing, as well as educating people about waste reduction and the benefits of recycling. These initiatives also demonstrate that recycling is a potential industry for development in Central Asia. The eco-activists use various techniques, from social awareness week programmes to the use of social media and mobile phone apps, to familiarise the public with the issue of waste management and its implications for the environment. These initiatives are indicative of societal shifts and greater preparedness to tackle more prominent environmental issues. While governments still have to take big steps towards reducing their carbon footprints and improving air and water quality, at the local level eco-entrepreneurs and activists are already contributing to this by taking action.

This chapter presented the local initiatives that demonstrate that waste collection, recycling and reduction are possible and doable in Central Asia. Moreover, with educational programmes to raise awareness about the importance of recycling, the correct ways of doing it, and the location of recycling points, the local initiatives are contributing to behavioural change and establishing the heading for a more environmentally conscious Central Asia.

References

Barvinskaya M (2019) Чиновники мэрии Бишкека так и не начали строить новую свалку. Прокуратура возбудила на них уголовное дело. Kloop. https://kloop.kg/blog/2019/01/11/chinovniki-merii-bishkeka-tak-i-ne-nachali-stroit-novuyu-svalku-prokuratura-vozbudila-na-nih-ugolovnoe-delo/. Accessed 28 Feb 2022

Decree of the President of the Republic of Kazakhstan (2013) On the concept for transition to "green economy" of the Republic of Kazakhstan. https://adilet.zan.kz/rus/docs/U1300000577. Accessed 2 Aug 2022

Eurasianet (2021) Film explores a dirty problem in Kyrgyzstan and Uzbekistan . https://eurasianet.org/film-explores-a-dirty-problem-in-kyrgyzstan-and-uzbekistan. Accessed 1 Feb 2022

European Bank for Reconstruction and Development (2013) Bishkek solid waste. https://www.ebrd.com/work-with-us/projects/psd/bishkek-solid-waste.html. Accessed 28 Feb 2022

Hashar (2020) Hasharweek RU. http://hasharweek.uz/#program. Accessed 2 Aug 2022

Independent Eco-Expertise Processes in Kyrgyzstan—comprehensive analysis of the current situation of the management system of solid waste. http://eco-expertise.org/sovershenstvovanie-ekologicheskoj-pol/stranovye/. Accessed 2 Aug 2022

Karimov N (2021) Bury and forget: how people "struggle" with growth of landfills in Tajikistan. Cabar. https://cabar.asia/en/bury-and-forget-how-people-struggle-with-growth-of-landfills-in-tajikistan. Accessed 28 Feb 2022

Kukula K (2018) Cleaning up Tajikistan's second largest city. https://www.ebrd.com/news/2018/cleaning-up-tajikistans-second-largest-city-.html. Accessed 1 Feb 2022

KWR (2022) Projects. https://www.kwr.kz/en/projects/. Accessed 28 Feb 2022

Majilisi Oli (2011) Law of the Republic of Tajikistan on industrial waste and consumption

Ministry of Ecology Geology and Natural Resources of the R of K (2021) Environmental code of the Republic of Kazakhstan. https://www.gov.kz/memleket/entities/ecogeo/documents/details/adilet/K2100000400?lang=ru. Accessed 2 Aug 2022

President of the Republic of Uzbekistan (2020) Decree of the President of the Republic of Uzbekistan on measures to improve activities in the field of handling with household and construction waste in the city of Tashkent. https://nrm.uz/contentf?doc=645282_postanovlenie_prezidenta_respubliki_uzbekistan_ot_15_12_2020_g_n_pp-4925_o_merah_po_sovershenstvovaniyu_deyatelnosti_v_sfere_obrashcheniya_s_bytovymi_i_stroitelnymi_othodami_v_gorode_tashkente&products=1_vse_zakonodatels. Accessed 2 Aug 2022

Rickleton C (2010) Kyrgyzstan: Bishkek confronts a waste management dilemma. Eurasianet. https://eurasianet.org/kyrgyzstan-bishkek-confronts-a-waste-management-dilemma. Accessed 28 Feb 2022

UNDP in Turkmenistan (2016) Sustainable cities in Turkmenistan: Integrated green urban development in Ashgabat and Awaza. https://www.tm.undp.org/content/turkmenistan/en/home/projects/sustainable-cities.html. Accessed 28 Feb 2022

United National Development Programme (2021) UNDP assists with the development of the Waste Management Strategy | UNDP in Turkmenistan. https://www.tm.undp.org/content/turkmenistan/en/home/presscenter/pressreleases/2021/undp-assists-with-the-waste-management-strategy.html. Accessed 28 Feb 2022

United Nations Development Programme (2021) Let's clean Kyrgyzstan's waste—the results of UNDP's eco-competition announced. https://www.undp.org/kyrgyzstan/news/lets-clean-kyrgyzstan's-waste-results-undp's-eco-competition-announced. Accessed 2 Aug 2022

United Nations Environment Programme (2017) Waste management outlook for Central Asia

United Nations Environment Programme (2019) Tajikistan: taking strides in chemicals and waste management. https://www.unep.org/news-and-stories/story/tajikistan-taking-strides-chemicals-and-waste-management. Accessed 28 Feb 2022

University of Central Asia (2019) UCA and USAID partner to improve reading skills in Tajikistan. https://ucentralasia.org/news/2019/october/uca-and-usaid-partner-to-improve-reading-skills-in-tajikistan. Accessed 2 Aug 2022

Usov A (2020) EBRD, EU and EIB help improve solid waste services in Kyrgyz Republic. https://www.ebrd.com/news/2020/ebrd-eu-and-eib-help-improve-solid-waste-services-in-kyrgyz-republic-.html. Accessed 28 Feb 2022

UZDaily (2020) Uzbekistan introduces a system of separate waste collection. https://www.uzdaily.uz/ru/post/57782. Accessed 2 Aug 2022

Vakulchuk R, Daloz AS, Overland I, Sagbakken HF, Standal K (2022) A void in Central Asia research: climate change. Cent Asian Surv: 1–20. https://doi.org/10.1080/02634937.2022.2059447

Vorotnikov V (2018) Can Kyrgyzstan sort it out? Recycling Waste World. https://www.recyclingwasteworld.co.uk/in-depth-article/can-kyrgyzstan-sort-it-out/115704/. Accessed 23 May 2022

Weinthal E, Watters K (2010) Transnational environmental activism in central Asia: the coupling of domestic law and international conventions. Env Polit 19:782–807. https://doi.org/10.1080/09644016.2010.508311

Wood C (2019) Bishkek's bright bins: recycling comes to the Kyrgyz capital. The Diplomat. https://thediplomat.com/2019/07/bishkeks-bright-bins-recycling-comes-to-the-kyrgyz-capital/. Accessed 1 Feb 2022

Ybyshova K (2022) Uzbekistan confronts its environmental challenges. International Finance Corporation. https://www.ifc.org/wps/wcm/connect/news_ext_content/ifc_external_corporate_site/news+and+events/news/insights/uzbekistan-confronts-its-environmental-challenges?cid=IFC_TT_IFC_EN_EXT. Accessed 2 Aug 2022

(2021) Astana residents are trained in separate waste collection. Kazakhstanskaya pravda. https://kazpravda.kz/n/astanchan-obuchayut-razdelnomu-sboru-othodov/. Accessed 2 Aug 2022

(2022) Fluid—biogas technologies http://www.fluid-biogas.com/. Accessed 2 Aug 2022

Aliya Tskhay is a Research Fellow at the School of Management at the University of St Andrews, United Kingdom. She holds a PhD in International Relations from the University of St Andrews. Dr Tskhay's research focuses on the study of the global energy sector with the focus on the net-zero targets, carbon-negative technologies, and energy transition. She also studies regional integration processes in Eurasia and provides extensive commentaries on the international cooperation in the region.

Open Access This chapter is licensed under the terms of the Creative Commons Attribution 4.0 International License (http://creativecommons.org/licenses/by/4.0/), which permits use, sharing, adaptation, distribution and reproduction in any medium or format, as long as you give appropriate credit to the original author(s) and the source, provide a link to the Creative Commons license and indicate if changes were made.

The images or other third party material in this chapter are included in the chapter's Creative Commons license, unless indicated otherwise in a credit line to the material. If material is not included in the chapter's Creative Commons license and your intended use is not permitted by statutory regulation or exceeds the permitted use, you will need to obtain permission directly from the copyright holder.

The manufacturer's authorised representative in the EU is Springer Nature Customer Service Centre GmbH, Europaplatz 3, 69115 Heidelberg, Germany. If you have any concerns regarding our products, please contact ProductSafety@springernature.com

Printed and bound by CPI Group (UK) Ltd, Croydon, CR0 4YY
25/03/2026
02078170-0017